8TH GRADE SCIENCE

Unit 5

Weather

AF876482

Table of Contents

CHARACTER SPOTLIGHT

DETERMINED

Fixing my mind and heart on the successful completion of a goal or purpose

OBJECTIVES

- Identify the characteristics of weather.
- Identify what causes wind.
- Explain how rotation impacts wind.
- Describe global winds and local winds.
- Describe surface currents and deep currents.
- Identify the impact of convection currents.
- Describe the factors that make up weather.
- Identify how solar energy impacts weather.
- Describe weather systems.
- Identify how weather systems change weather.
- Explain how hurricanes, tropical cyclones, and typhoons develop.
- Describe the scale system and hurricane damage.
- Identify the difference between climate and weather.
- Describe the greenhouse effect and global warming.
- Understand the different factors that influence climate.
- Understand the main factors that influence climate change.
- List the evidence for a warmer earth.
- Describe both natural and human factors for climate change.
- Understand possible future consequences of climate change.

CHARACTER SPOTLIGHT

What would you like to accomplish? Is it a right and proper thing to do? *What would prevent you from doing it?*

Many people have a list of things they would like to do. Unfortunately, most people who have these lists never accomplish much of what is on them. That is not because of a lack of desire. It is because of a lack of determination. Determination requires resolve, planning, and work. In life, very few things happen by accident. Things happen because someone has learned to plan their work and work their plan.

SECTION ONE

Objectives:

- Identify the characteristics of weather.
- Identify what causes wind.
- Explain how rotation impacts wind.
- Describe global winds and local winds.
- Describe surface currents and deep currents.
- Identify the impact of convection currents.

Vocabulary:

air mass – air that covers a large area of land or ocean with the same temperature and humidity

air pressure – the weight of the air on the surface of the earth

Coriolis *[kor-ee-OH-lis]* **effect** – the curving path of an object due to the earth's rotation

current – the movement of water

front – the boundary where two air masses meet

salinity *[suh-LIN-uh-tee]* – the amount of salts or solids dissolved in a liquid

weather – the state of the atmosphere at a given time and place

1. WEATHER AND WIND

When you wake up in the morning, you might check the weather to help you choose how to dress for the day. Weather has a major impact on choice of clothing, transportation, and schedule of events. It can even impact a person's mood and feelings. **Weather** is the state of the atmosphere at a given time and place. Weather is formed through a complex relationship between water (hydrosphere) and air (atmosphere). The two systems collide and intertwine to form thunderstorms, rain, snow, winds, and hurricanes.

Weather changes frequently and is dependent on various factors. The main factors affecting weather are day-to-day temperature, humidity, precipitation, air pressure, and wind.

- Temperature is how hot or cold the atmosphere is and is measured with a thermometer. Temperature is reported in Fahrenheit in the United States and in Celsius in the rest of the world and by scientists.
- Humidity is the amount of water vapor found in the air. Clouds, rain, and snow are all formed by the amount of water vapor in the atmosphere. Humidity is displayed as the percentage of maximum amount of water that air can hold, or relative humidity. Cold air holds less water than warm air. When the relative humidity is at 100%, the air is saturated with water vapor. Once saturation occurs, the extra water vapor is released as precipitation.
- Driven by energy from the sun, surface waters (hydrosphere) in lakes, streams, rivers, and oceans evaporate into the atmosphere, condense, and fall back to the ground as precipitation. Precipitation can be rain, snow, sleet, or hail. While increased precipitation can replenish water supplies and support agriculture, too much precipitation can lead to flooding. However, too little precipitation over an extended period of time can result in drought-like conditions in a particular area or region.
- **Air pressure** is the weight of the air on the surface of the earth. Changes in the amount of pressure impact weather systems, changing sunny days into rainy ones. Pressure is measured in atmospheres.
- Wind is the movement of air and forms due to the differences in temperature and pressure.

With the sun as the driver, the hydrosphere and the atmosphere both interact to influence weather and climate here on our planet. In this Lesson, we will focus on the causes of weather and the impact of wind on weather. Later in this Unit, we will look at how oceans work to affect weather and the global climate.

What Causes Weather?

The earth's atmosphere is divided up into different layers based on temperature and pressure. When the composition and layers of the atmosphere converge with the water cycle, or hydrosphere, weather is created. Nearly all weather occurs in the lowest layer of the atmosphere closest to the surface of the planet, in a region known as the troposphere. The troposphere is the innermost layer of the earth's atmosphere. It contains the air we breathe and the clouds in the sky. The heat from the sun causes the air in this lower layer to warm to different temperatures at different locations around the globe. When warm air rises, cold air rushes in from below to replace it. Wind is the term used to describe this movement of air. Changes in the weather, such as clear, sunny skies or heavy rain, are brought on by winds.

Thunderclouds produce hail, thunder, and lightning.

The humidity or amount of water vapor (gas) in the air also affects the weather. The level of moisture in the air influences how clouds form and the weather they produce. Clouds in the atmosphere are vaporized water. They differ in size, shape, and color. Certain types of clouds are associated with certain types of weather. Some clouds bring heavy precipitation, like heavy rain or snow, while the precipitation from some clouds might fall to the ground as a drizzle or light snow.

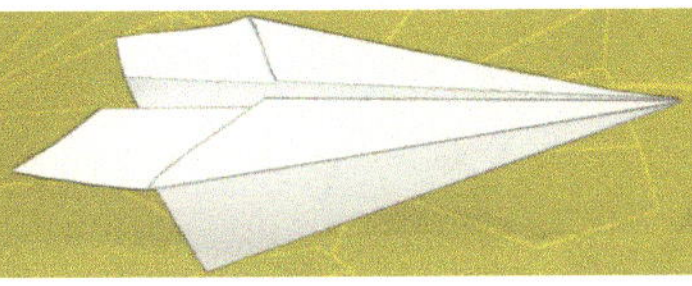

REVIEW

Match the words with the descriptions.

1.1) _____ temperature

1.2) _____ air pressure

1.3) _____ humidity

1.4) _____ troposphere

1.5) _____ wind

1.6) _____ precipitation

1.7) _____ weather

1.8) _____ atmospheres

A. rain, snow, hail, or sleet

B. the movement of air

C. the lowest layer of the atmosphere, closest to the earth's surface

D. how hot or cold the atmosphere is

E. the amount of water vapor in the air

F. the weight of the air on the surface of the earth

G. what pressure is measured in

H. the state of the atmosphere at a given time and place

Choose the best answers.

1.9) _____ Which answer best describes weather?
- A. the amount of rain a specific area receives
- B. the temperature of the air
- C. the condition of the atmosphere at a given time and place
- D. the average weather in a certain area over an extended period of time

1.10) _____ In which layer of the atmosphere does most weather on the earth occur?
- A. troposphere
- B. mesosphere
- C. stratosphere
- D. thermosphere

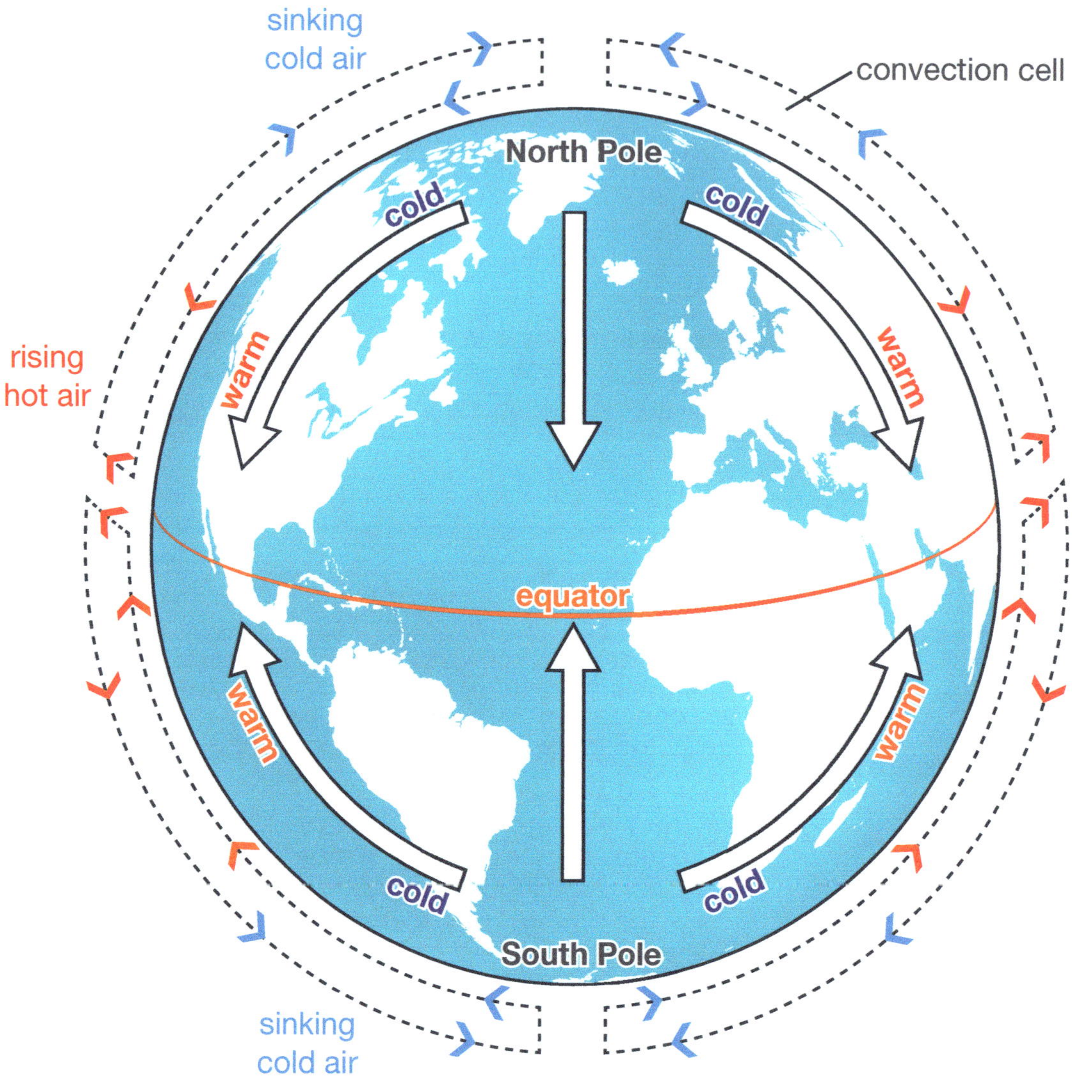

Fig. 1.1 Global convection cell patterns

WHAT CAUSES WIND?

Whenever you feel a breeze or a gust of wind, you are feeling the movement of air caused by differences in air pressure. Air can have different densities. These differences occur due to the uneven heating of the earth's surface from the sun. For example, land (rock) heats and cools faster than water (ocean). Uneven heating causes different temperatures. Different temperatures create different densities in the air.

The differences in temperature and density cause air to move through convection. Convection is the movement of matter due to differences in density. Cold air is denser than warm air, so it sinks, causing greater pressure on the earth. This area of greater pressure is called high air pressure. The air then moves from these areas of high air pressure toward the areas of low air pressure, causing wind. The larger the difference in air pressure, the faster the movement of air or wind speed. The energy from the sun drives convection in the earth's atmosphere.

Areas of high air pressure can be found at the North and South Poles. Areas of low air pressure are located at the equator where there is warmer and less dense air. The air moves in convection cells, or small circular patterns, that allow the air to move from the equator, cool, and become denser (**Fig. 1.1**). This pattern of movement creates belts of high pressure at 30° N and 30° S latitudes from sinking air. Close to the poles at 60° N and 60° S latitudes, low-pressure belts form when the warm air is pushed up.

Another factor that plays into wind patterns is the rotation of the earth. The earth rotates on an axis, which causes the winds to curve around it. If the earth did not rotate, then winds would blow in a straight line. Imagine a person standing still at the equator and another person standing still at the North Pole. As the earth makes a complete rotation in one day, the person standing at the equator will travel a longer distance than the person at the North Pole. The person at the equator also travels slightly faster than the person at the North Pole. As air moves from the equator to the North Pole, it keeps the same speed and direction. This difference in rotation and speed causes air toward the north and south of the equator to flow in different directions. Air moving north of the equator moves clockwise and air moving to the south moves counterclockwise. These patterns are called the Coriolis effect. The **Coriolis effect** is the curving path of an object due to the earth's rotation (**Fig. 1.2**).

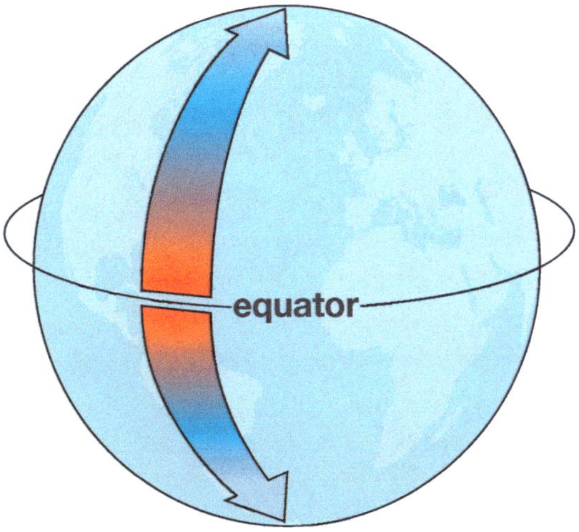

Fig. 1.2 The Coriolis effect causes wind to move in a curved path.

Global Winds

The pressure belts and the Coriolis effect make patterns of calm air and more active wind systems. The wind systems located at the earth's surface are called global winds. There are several different types of winds named by their locations and the directions they flow.

Trade Winds

Trade winds blow between the equator and 30° latitudes in both the Northern and Southern Hemispheres. Due to the earth's rotation, trade winds curve to the west, which means Northern Hemisphere trade winds come from the northeast, and Southern Hemisphere trade winds come from the southeast. Trade winds got their name from sailors who used these winds to sail from Europe to the Americas.

Westerlies

Westerlies blow between 30°–60° latitudes in both the Northern and Southern Hemispheres. Westerlies curve to the east; therefore, westerlies in the north come from the southwest, and westerlies in the south come from the northwest. Westerlies carry moist air over the United States and produce snow and rain.

Polar Easterlies

Polar easterlies blow between 60° latitudes and the North and South Poles. Polar easterlies come from cold, sinking air that moves from the poles toward 60° N and 60° S latitudes and curve to the west. At the North Pole, polar easterlies carry cold, arctic air over the United States and produce snow and freezing conditions.

Doldrums

Doldrums are a calm area where trade winds from the Northern and Southern Hemispheres meet. This area remains calm because there is warmer and less dense air that creates an area of low pressure.

Horse Latitudes

Near the 30° latitudes, the air stops moving and sinks, forming a calm area called the horse latitudes. The name may have originated from when ships carrying horses from Europe to the Americas would get stuck in these areas due to the lack of wind. Since fresh water was limited aboard the ships, the sailors could not afford to wait for the winds to pick up. Horses were thrown overboard to lighten the load and get the ship moving sooner.

Jet Streams

Areas of high-speed winds blowing west to east are called jet streams. Jet streams are narrow belts of winds that blow about 7–16 km above the earth's surface. Airplanes can take advantage of these jet streams and fly from the West Coast to the East Coast at a faster pace than other planes traveling in the opposite direction. The polar jet stream is located in the Northern Hemisphere, while the subtropical jet stream is located in the Southern Hemisphere. Jet streams form between areas of hot and cold air and can shift north and south. During the wintertime when temperatures are cooler, the polar jet stream will move south and bring cold, arctic air into the United States. In the spring when the temperature rises, the jet stream shifts to the north.

Local Winds

The geographic features of an area can influence local winds. The location of bodies of water or a mountain can create differences in temperature and pressure that create these local winds. While global winds move over a long distance, local winds move over short distances and can blow from any direction depending on the features of that area.

Large bodies of water such as the ocean or a lake take longer to warm up than land does. The air above the land becomes warmer than the air above the water. As the colder air from the water moves over the land, the warm air pushes up, creating a sea breeze. At night, the air over the land will cool faster than the air over the water, creating a high-pressure zone over the land. Wind that blows from the land toward the water is called a land breeze (**Fig. 1.3**).

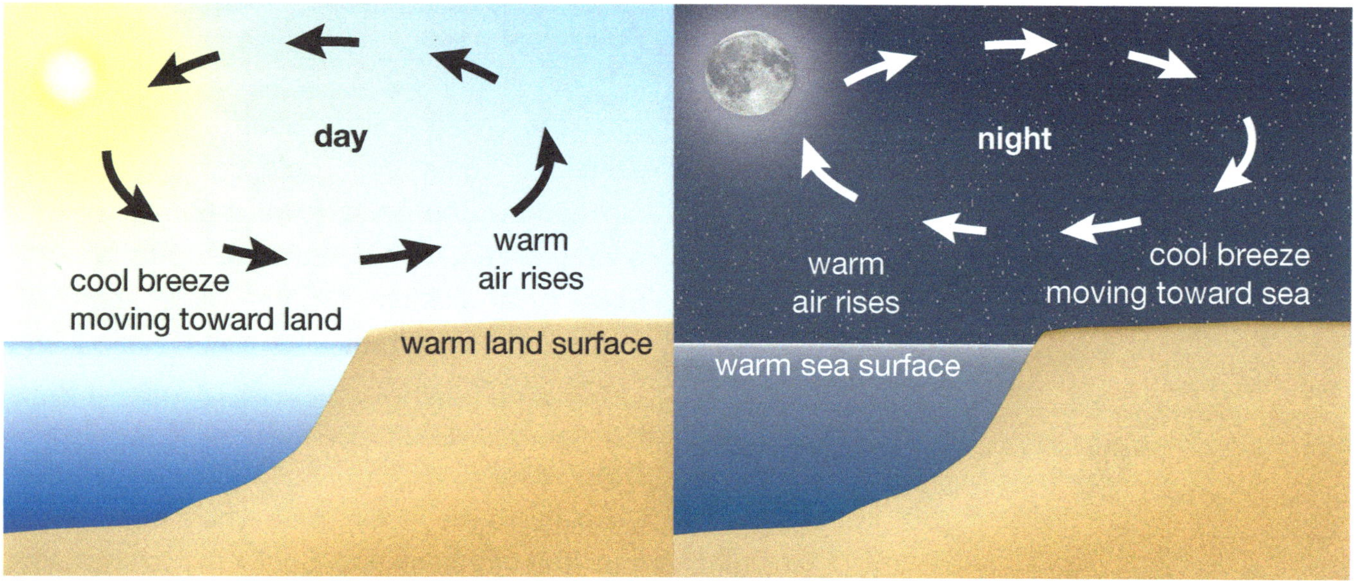

Fig. 1.3 Sea and land breezes

Mountains and valleys create mountain and valley breezes (**Fig. 1.4**). During the daytime, the sun warms the air along the slopes of the mountain faster than the air in the valley. This creates areas of low pressure at the tops of the mountains, creating a valley breeze. A valley breeze moves from the valley up the slopes of the mountain. At nighttime, the air along the slopes cools and moves down the valley, creating a mountain breeze.

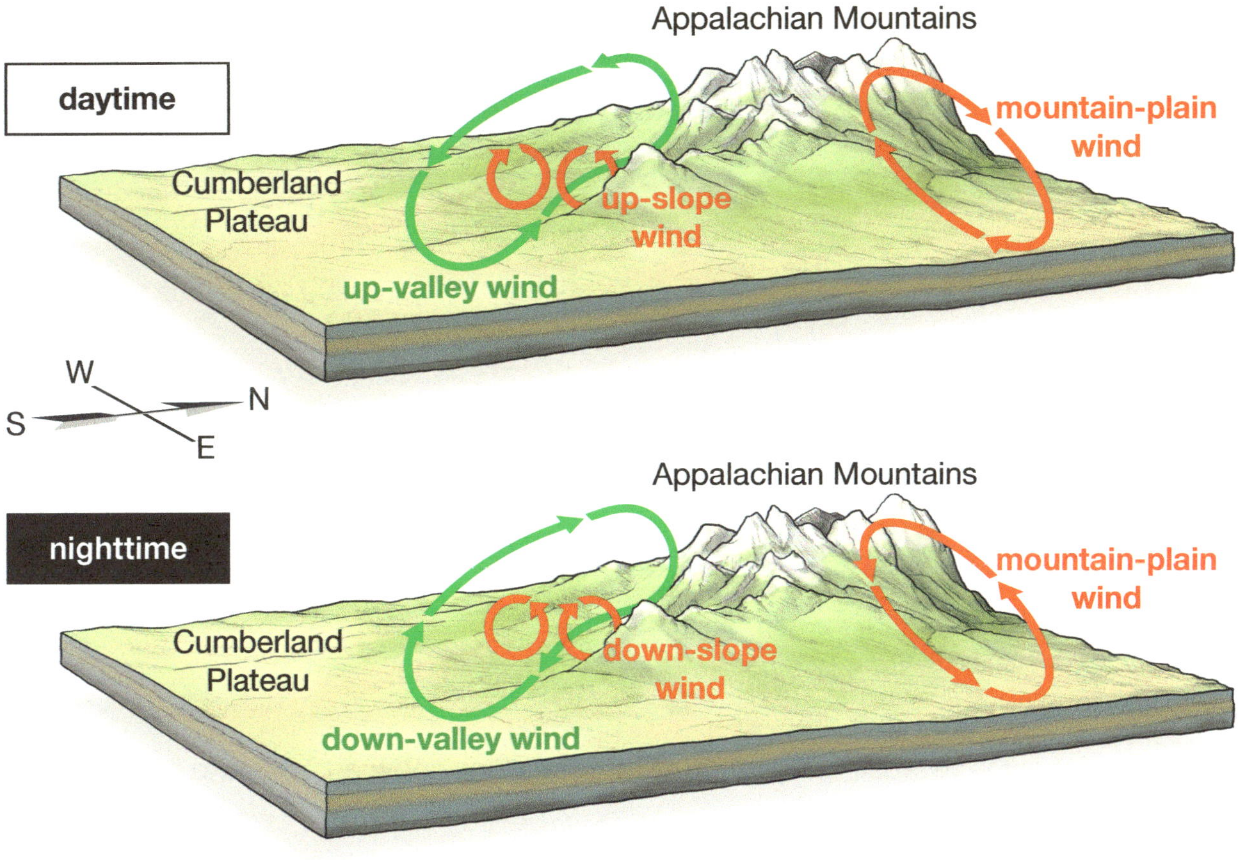

Fig. 1.4 Mountain and valley breezes

Underline the best answers.

Cold air is denser than warm air and 1.11) (**sinks**, **rises**)*, causing greater pressure on the earth. This area of greater pressure is called* 1.12) (**low**, **high**) *air pressure. The air then moves from these areas of high air pressure toward the areas of low pressure, causing wind. The* 1.13) (**smaller**, **larger**) *the difference in air pressure, the faster the movement of air or wind speed. The energy from the* 1.14) (**ocean**, **sun**) *drives the convection in the earth's atmosphere.*

Match the wind-related words with the descriptions.

1.15) _____ Coriolis effect

1.16) _____ trade winds

1.17) _____ westerlies

1.18) _____ polar easterlies

1.19) _____ doldrums

1.20) _____ horse latitudes

A. the calm area where air stops and sinks

B. the curve west between 60° latitude and the North and South Poles

C. the calm area where trade winds meet

D. the curve west between equator and 30° latitudes

E. the curve east between 30°–60° latitudes

F. the curving path an object takes due to the earth's rotation

Choose the best answers.

1.21) ________ carry moist air and produce snow and rain.

A. Westerlies
B. Polar jet streams
C. Polar easterlies
D. Jet streams

1.22) ________ curve to the west, carry arctic air, and produce snow and freezing conditions.

A. Westerlies
B. Polar jet streams
C. Polar easterlies
D. Jet streams

1.23) ________ move south and bring arctic air to the United States.

A. Westerlies
B. Polar jet streams
C. Polar easterlies
D. Jet streams

1.24) ________ are areas of high-speed winds blowing from west to east.

A. Westerlies
B. Polar jet streams
C. Polar easterlies
D. Jet streams

1.25) _____ Shifting winds that cause major rainstorms, flooding streams, lakes, and rivers are interactions between the ___.

A. atmosphere and geosphere
B. atmosphere and hydrosphere
C. biosphere and geosphere
D. hydrosphere and geosphere

1.26) The uneven heating of the earth's surface from the sun causes local and global winds. Which THREE items are examples of local winds?

- ☐ sea breezes
- ☐ trade winds
- ☐ polar easterlies
- ☐ mountain breezes
- ☐ jet streams
- ☐ valley breezes

CHECK ☐ CORRECT ☐ RECHECK ☐

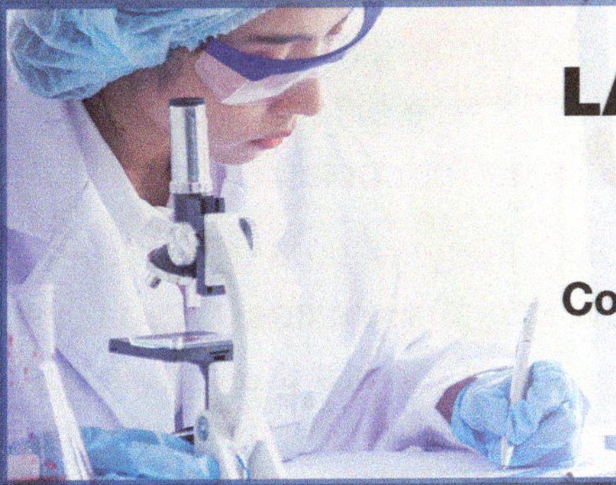

LAB: THE SUN'S EFFECT ON EARTH'S SURFACE

Complete this lab in the Lab Manual.

Teacher Initial ☐

CHARACTER SPOTLIGHT

"You can have anything you want if you want it badly enough. You can be anything you want to be, do anything you set out to accomplish if you hold to that desire with singleness of purpose."

– Abraham Lincoln

2. CURRENTS

Have you ever written a message in a bottle and thrown it into the sea? If not, perhaps you read about it in a book or saw it in a movie. The hope and expectation is that the bottle will be carried to another distant shore where a stranger will read the message.

How does a bottle, like the one in the previous example, move from one location to another? The answer is through currents. When referring to water, **currents** are the movements of water. Ocean currents are caused by the tides, winds, or the differences in density between cold and warm water. Currents have a great effect on weather, the transportation of goods over the seas, and the movement of nutrients.

Ocean currents function much like a conveyer belt. Warm water and precipitation move from the equator to the poles, and cold water moves from the poles back to the equator. Ocean currents regulate the global climate by balancing out the unequal distribution of solar energy that reaches the earth's surface. Regional temperatures would be more severe without ocean currents, with extremely hot temperatures near the equator and icy temperatures at the poles. Much of the land on the earth would be completely inhabitable.

There are two types of ocean currents: surface currents and deep currents.

Surface Currents

Surface currents only go about 100 meters deep into the ocean but are found across all oceans (**Fig. 2.1**). Surface currents and the pattern they move in are influenced by several factors: continental deflections, the Coriolis effect, global winds, and the sun.

- Continental deflections occur when surface currents run into continents. The collision of the water and the continent causes the current to change direction. For example, the South Equatorial currents run toward the west in the Atlantic Ocean but turn south once they run into South America. This shift in direction can cause different areas to become warmer or cooler.
- The Coriolis effect ensures that currents do not flow in a straight line. Ocean currents will flow toward the east in the Northern Hemisphere and toward the west in the Southern Hemisphere.
- Global winds push water across the surface of the earth, causing surface currents. The direction and different types of wind cause the currents to flow in different directions. For example, the winds at the equator blow east to west; therefore, the surface currents move in an east-to-west direction.
- The sun is responsible for providing the energy needed for surface currents. At the equator, the sun heats the air more than other latitudes away from the

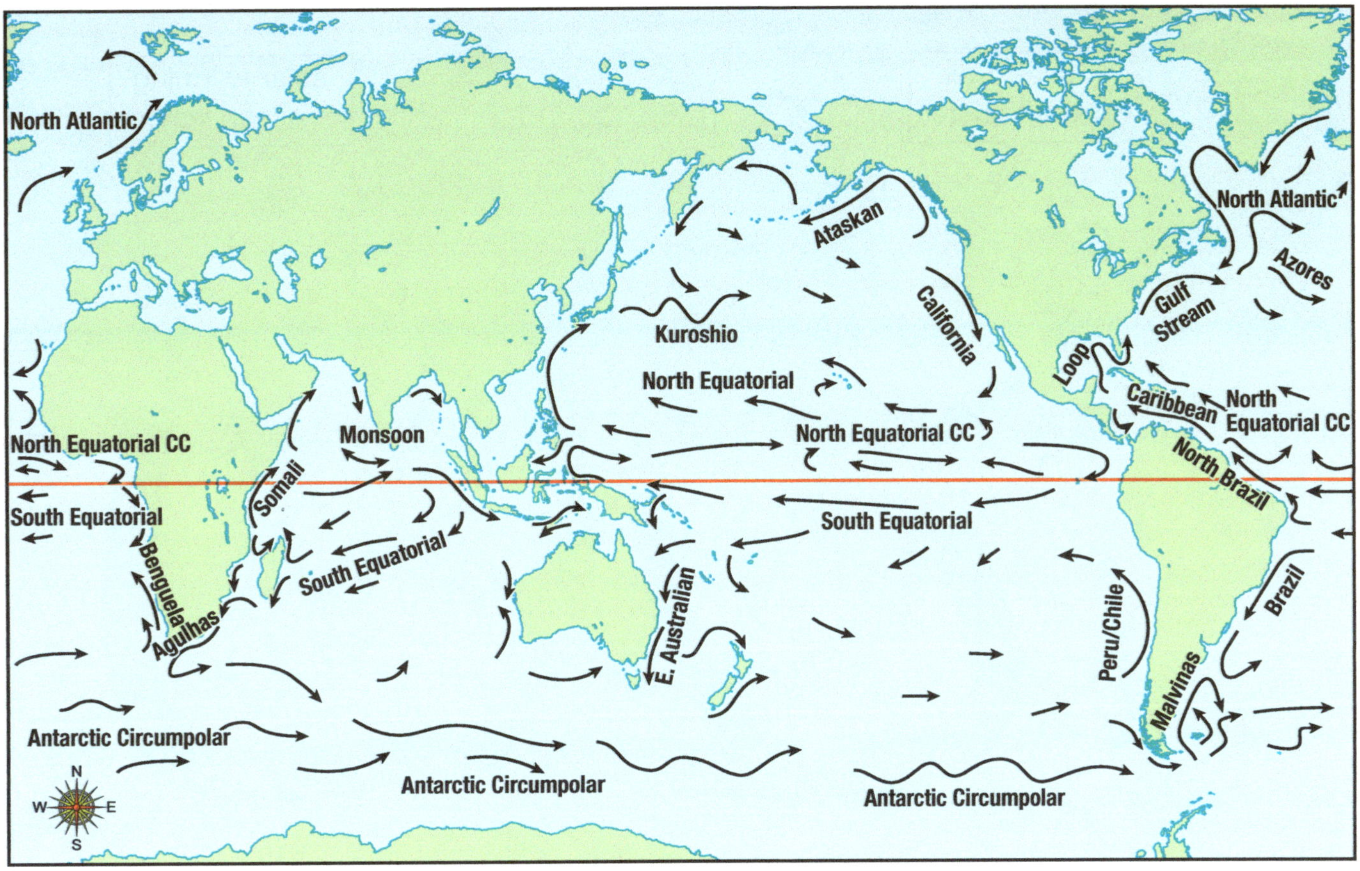

Fig. 2.1 Ocean surface currents

Source: Mountain High Maps® Copyright © 1993-2020 Digital Wisdom®, Inc.

equator. This heating causes pressure differences, creating a low-pressure area near the equator. When there are pressure differences, wind is created, which in turn causes surface currents to form.

The Gulf Stream is an example of a surface current. This current moves more water than all the rivers in the world combined. The Gulf Stream is responsible for moving warm water north, up the East Coast of the United States.

Deep Currents

Ocean currents moving deeper than 100 meters are called deep currents. Deep currents occur from differences in the density of the water. The density in the ocean is determined by the salinity and temperature of the water. **Salinity** is the amount of salts or solids dissolved in a liquid. Water that has a high salinity is denser than water with a low salinity. The temperature impacts density due to the movement of the molecules. In cold water, the molecules have less movement and are closer together, allowing more molecules to fit in a space. This makes cold water denser than warm water. In warm water, the molecules move faster and farther apart and the water becomes less dense. The less dense water is able to rise above the cold water. When the water at the surface of the ocean becomes denser, it sinks under the less dense water (**Fig. 2.2**). This movement creates deep currents. Deep currents move on top of denser water, or along the ocean floor. The densest and deepest water currents are found near Antarctica.

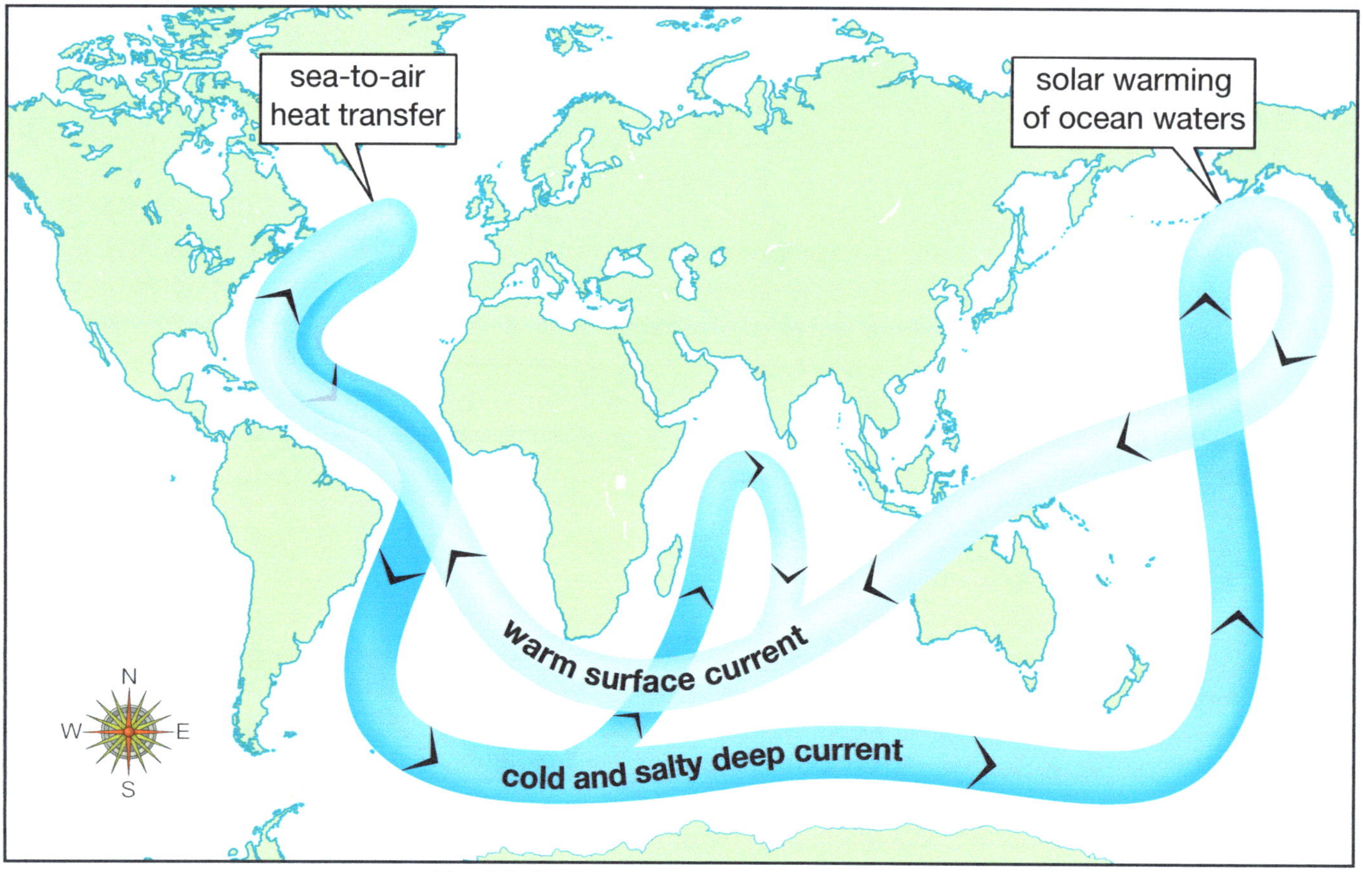

Fig. 2.2 Deep currents and surface currents

Source: Mountain High Maps® Copyright © 1993-2020 Digital Wisdom®, Inc.

Convection Currents

Surface currents and deep currents are linked in the ocean to form convection currents. Convection currents are the movements of water due to differences in the density of different water sources. Convection currents can move up and down, in a circle, or in a cycle. Surface currents run on the top while deep currents run on the bottom. As the density of the water changes, the surface current can become a deep current; or, deep current water can become surface water if it is less dense.

Convection currents are responsible for transferring energy from the sun to colder regions. Surface currents absorb energy from the sun. As the warm water loses energy to its surroundings, the warm water cools, becomes denser, sinks, and travels along the bottom of the ocean. Cold water at the bottom will then rise to the surface and absorb energy from the sun and continue the cycle. At the equator, a large amount of energy is absorbed from the sun and the atmosphere. The energy that is absorbed is moved toward the North and South Poles in the ocean currents. As warm water moves to cooler areas, the energy is released back into the atmosphere, which has a major impact on the earth's climate.

Transportation

Ocean currents also transport matter. Whatever matter is found in the ocean will move along with the ocean current. Dissolved solids such as salt are the most common particles moving in an ocean current. Other nutrients such as nitrogen and phosphorus are also transported in ocean currents and are important for plant and animal growth.

When surface winds blow the surface currents away from the shore, the warm surface water is replaced by cold water that is full of nutrients from the deep currents. This process is called upwelling and is responsible for bringing important nutrients such as nitrate and iron to the surface to support the growth of ocean life (**Fig. 2.3**). The areas where upwellings occur are locations where ocean animals can survive well and are often the location of fisheries. If an upwelling is reduced or disrupted, then ocean life will also be reduced.

Atmospheric gases that are absorbed into the ocean are transported using currents. Nitrogen, oxygen, argon, and carbon dioxide are the gases most commonly found in our atmosphere. Because of absorption, they are also found in the ocean.

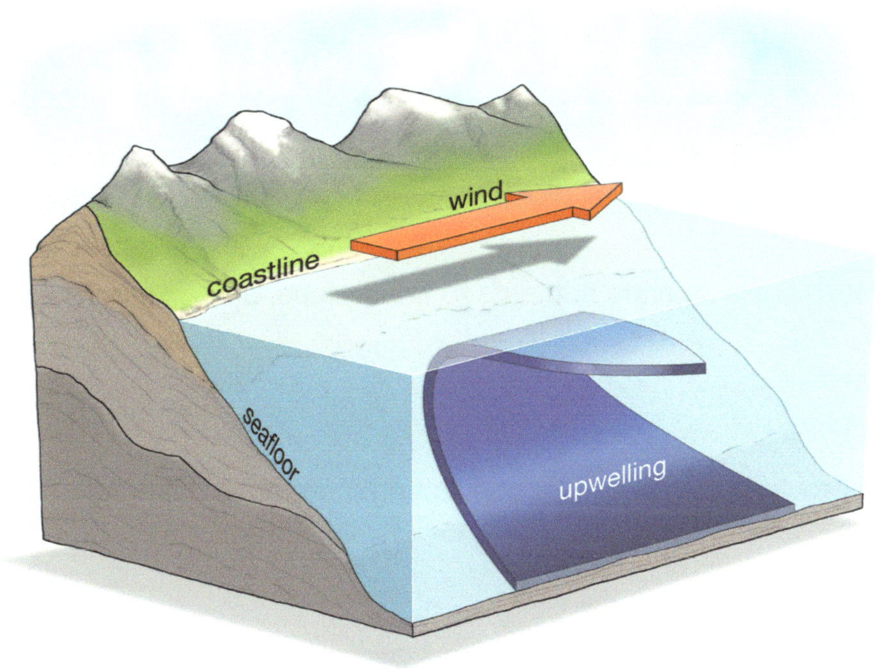

Fig. 2.3 Upwelling

REVIEW

Choose the best answers.

2.1) _____ The movement of ocean water is called an ocean ___.

A. current
B. convection
C. conduction

2.2) _____ Ocean currents that go no more than 100 meters deep in the ocean are called ___ currents.

A. convection
B. deep
C. surface

2.3) _____ Ocean currents that move deeper than 100 meters are called ___ currents.

A. high
B. middle
C. deep

2.4) _____ The movements of water due to the differences in density are called ___ currents.

A. convection
B. deep
C. surface

2.5) _____ As the ___ of the water changes, surface currents can become deep currents.

A. wildlife
B. density
C. distance

2.6) _____ When deep currents bring nutrients to the surface, a(n) ___ is developed.

A. upwelling
B. downwelling
C. transpiration

2.7) _____ The ___ provides the energy for convection currents in the ocean.

A. moon
B. sun
C. tides

2.8) Which FOUR items do ocean currents transport?

- ☐ energy
- ☐ global winds
- ☐ matter
- ☐ nutrients
- ☐ gases

Write M for More dense or L for Less dense.

2.9) _____ warm ocean water

2.10) _____ cold ocean water

2.11) _____ low salinity water

2.12) _____ high salinity water

CHECK | CORRECT | RECHECK

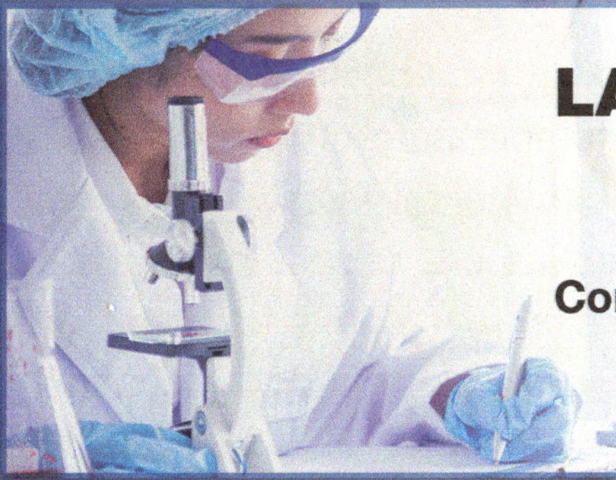

LAB: CURRENT DENSITY AND SALINITY

Complete this lab in the Lab Manual.

Teacher Initial

CHARACTER SPOTLIGHT

DETERMINATION IS CONNECTED

Determination involves many other character traits. It is not just desiring and deciding. Rather, it requires so much more:

- Optimism to believe that your efforts definitely will pay off
- Studiousness to learn the things you will need to know in order to succeed
- Persistence and Patience because the process may be very long
- Hopefulness to keep you going during the tough times

3. ATMOSPHERIC MOVEMENT

Much like two ancient armies clashing in battle, turbulent weather is found where two different air masses meet in the atmosphere.

AIR MASSES

An **air mass** is air that covers a large area of land or ocean with the same temperature and humidity. Air masses are so large, they might be 1,000 miles across and reach several miles into the atmosphere. Air masses are constantly in motion and move in certain directions depending on their location. In the U.S., most air masses move west to east. Air masses vary in temperature and can be continental (land) or maritime (ocean) in origin.

The amount of moisture in an air mass determines what type of air mass it is. Continental air masses have a low moisture content, making them more dry. They are represented by a lowercase "c" and begin over continents. Maritime air masses begin over oceans and are more moist. They are represented by a lowercase "m."

Continental and maritime air masses are further divided into separate masses based on temperature and the place they first form (**Fig. 3.1**). Arctic air masses, represented by an uppercase "A," are cold masses that begin over the arctic or antarctic regions. Polar air masses, represented by an uppercase "P," begin over high latitudes over land or sea. They are cold but not as cold as arctic air masses. Tropical air masses are represented with an uppercase "T" and are warm or hot masses that come from low latitudes over land or sea.

Putting two letters together describes the location and temperature of an air mass. For example, a continental arctic mass would be represented as cA, describing that it is a continental mass that came from the arctic region and is very cold and dry. A continental polar mass, or cP, would also be a dry mass that is not as cold as the cA. A maritime polar mass, mP, would be a cold mass that originated over the ocean; therefore, it

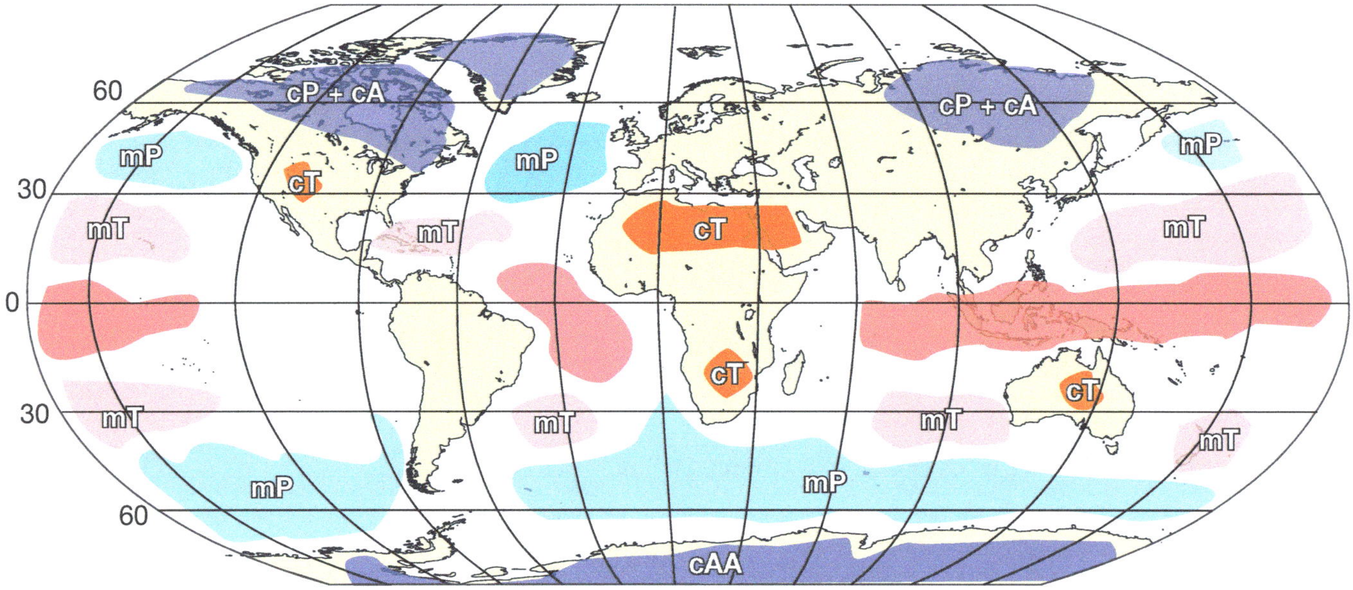

Fig. 3.1 Global air masses

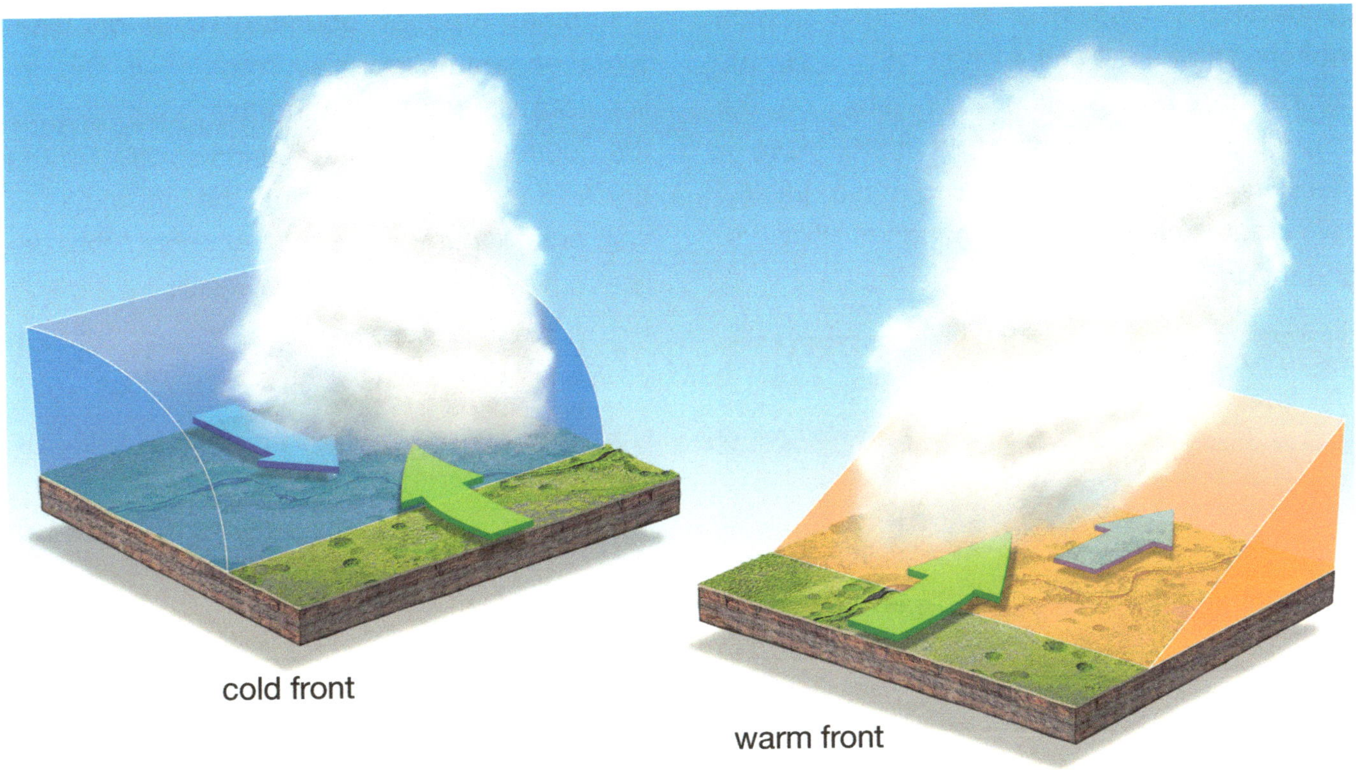

Fig. 3.2 Cold front and warm front

would have a high moisture content. An air mass over a desert region would be cT, as it is continental and tropical, meaning hot and dry.

As air masses begin to move around the earth, they collect local characteristics and change temperature and moisture content. For example, an arctic mass that moves over the ocean will pick up warmth and moisture and change into a maritime polar mass, or mP. Now, the original air mass is not as cold and contains moisture. If that air mass moves south from the arctic and picks up warmth from the ground, it will dry out and become a continental polar air mass, cP.

Air Mass Boundaries

Sometimes, two air masses will meet. When this happens, it is called a front. **Fronts** were named by scientists from Switzerland. A front reminded them of a battle line between two armies when they meet each other. When two air masses meet, the air does not mix. Rain and storms often happen at a front.

There are two different types of fronts: cold and warm (**Fig. 3.2**). Each type of front brings a different type of weather. A cold front occurs when a cold air mass moves over a warm air mass. A warm front occurs when a warm air mass moves over a cold air mass. The air in a cold air mass is denser than the air in a warm air mass. As the colder air mass moves under the warmer air mass, the warm air moves up over the cold air, rising into the atmosphere. If the boundary between two air masses does not move, a stationary boundary develops.

Weather is impacted by the motion and structure of air masses. Fronts have a vertical structure with a slope. Warm fronts have a gentle slope, causing air to rise in a gradual motion. The gentle slope often causes stratus clouds and precipitation to develop on the north side of the front. Cold fronts have a steep increase that forces air to move quickly. This creates clouds, narrow bands of showers, and thunderstorms ahead of the front.

Another type of boundary is a dry line, which separates moist air from dry air. The dry line normally forms from north to south across the central and southern plains states. As the dry line moves east during the afternoon, it can create thunderstorms. At night, the dry line moves back toward the west.

Stratus (layered) clouds often form at warm fronts while thunderstorms form at cold fronts.

REVIEW

Choose the best answers.

3.1) _____ A cold and dry air mass that starts over land is a(n) ___.
A. tropical air mass
B. arctic air mass
C. maritime air mass
D. continental air mass

3.2) _____ A moist air mass that starts over an ocean is a(n) ___.
A. tropical air mass
B. arctic air mass
C. maritime air mass
D. continental air mass

3.3) _____ Cold air masses that begin over arctic areas are ___.
A. tropical air masses
B. arctic air masses
C. polar air masses
D. continental air masses

3.4) _____ Cold air masses that begin over high latitudes from land or sea are ___.
A. tropical air masses
B. arctic air masses
C. polar air masses
D. continental air masses

3.5) _____ Warm air masses that come from low latitudes over land or sea are ___.

A. tropical air masses
B. arctic air masses
C. polar air masses
D. continental air masses

3.6) _____ A ___ occurs when a cold air mass moves over a warm air mass.

A. front
B. hot front
C. warm front
D. cold front

3.7) _____ A ___ occurs when a warm air mass moves over a cold air mass.

A. dry line
B. hot front
C. warm front
D. cold front

3.8) _____ The ___ separates moist air from dry air.

A. dry line
B. hot front
C. warm front
D. cold front

AIR PRESSURE SYSTEMS

Weather is impacted by air pressure. The amount of pressure on the surface of the earth changes from day to day since the earth is not heated equally by the sun. Areas with warmer air have lower air pressure because the warmer air is rising. Colder air creates a high air pressure system.

LOW AIR PRESSURE SYSTEM

A low air pressure system has lower air pressure at the center of an air mass. Storms, precipitation, and other "bad" weather form in areas of low pressure. As the wind blows toward the low pressure, the air in the atmosphere rises. When the air rises, the water vapor in the atmosphere condenses and forms clouds and precipitation. Low-pressure systems, indicated by a red uppercase "L," have whirling masses of warm, moist air that create stormy weather with strong winds. The winds in a low-pressure system move counterclockwise north of the equator and move clockwise when located south of the equator. This movement is called cyclonic flow and occurs due to the earth's spin and the Coriolis effect (**Fig. 3.3**).

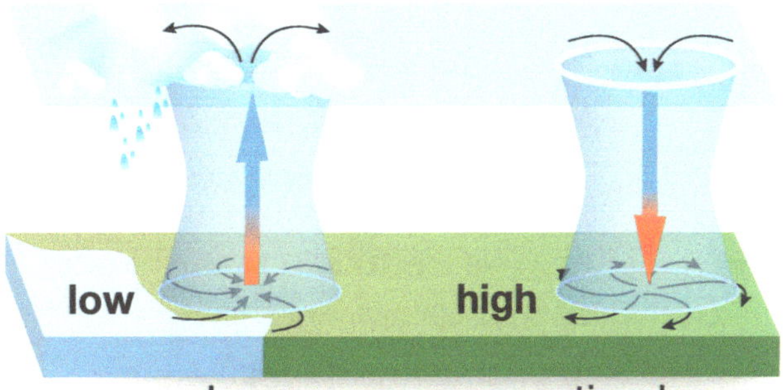

Fig. 3.3 Cyclonic flow and anticyclonic flow

HIGH AIR PRESSURE SYSTEM

A high air pressure system has higher air pressure at the center of an air mass. High-pressure systems, indicated by a blue uppercase "H," contain a whirling mass of cool, dry air, creating fair weather, light winds, and sunny skies. The winds blow away from a high-pressure system in the opposite direction of a low-pressure system. The winds move clockwise north of the equator and counterclockwise south of the equator. This air movement is called anticyclonic flow (**Fig. 3.3**), as the air from high in the atmosphere sinks down and fills in space from the air that is moving out from the system.

Air pressure systems are measured using electronic sensors, or barometers. Sensors are able to measure the amount of pressure continuously over time. Barometers are used to measure how much air is being pushed from the system and are often used by meteorologists today.

Barometer

Choose the best answers.

3.9) _____ A ___ is a whirling mass of warm air that creates stormy weather.
A. low air pressure system
B. high air pressure system
C. warm front
D. cold front

3.10) _____ A ___ is a whirling mass of cool, dry air that creates sunny skies.
A. low air pressure system
B. high air pressure system
C. warm front
D. cold front

Write H for High air pressure or L for Low air pressure.

3.11) _____ stormy weather

3.12) _____ sunny skies

3.13) _____ clockwise north of the equator

3.14) _____ counterclockwise north of the equator

3.15) _____ cyclonic movement

3.16) _____ anticyclonic movement

CHECK ☐ CORRECT ☐ RECHECK ☐

ACTIVITY: High and Low Pressure

Materials:

- large sponge

Directions:

1 – Press down on or apply pressure to the large sponge.

2 – Observe what happens.

3 – Release the pressure on the sponge after a few seconds.

4 – Observe what happens.

5 – Do this several times.

6 – Use the image of high and low air pressure to answer the questions.

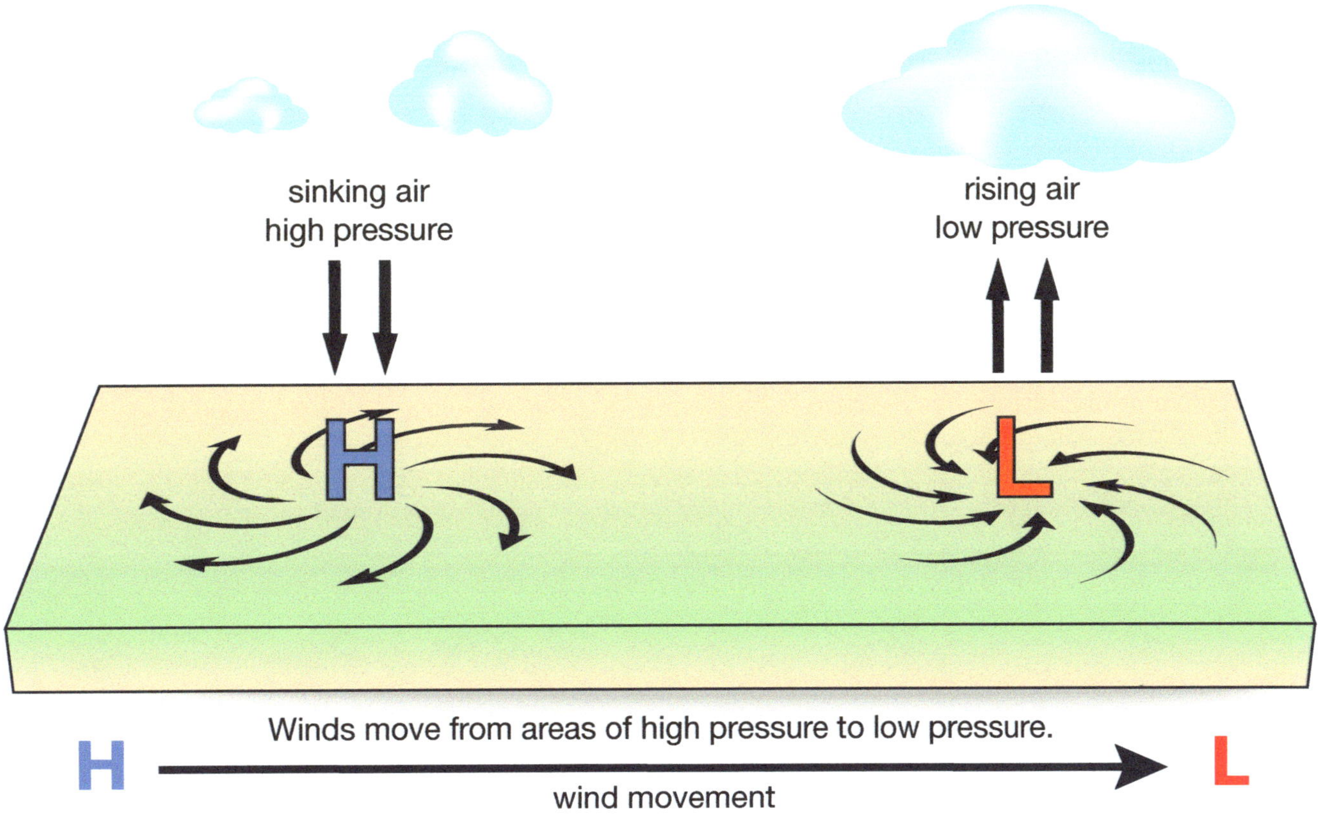

Write the correct answers.

1) Describe how the sponge activity relates to the image of high and low air pressure systems.

____

____

____

2) Where would you expect to find sunny weather and why? ______________________________

3) Where would there be cloudy or stormy weather and why? ______________________________

4) What is happening to the air in a high-pressure system? ______________________________

5) What is happening to the air in a low-pressure system? ______________________________

6) Wind moves from areas of ______________ pressure to ______________ pressure.

TEACHER CHECK

MY SCORE

QUIZ 1

TEACHER INITIAL

(Each answer, 4 points)

Match the descriptions with the words.

1.01) _____ the movement of air

1.02) _____ the curving path an object takes due to the earth's rotation

1.03) _____ the amount of water vapor in the air

1.04) _____ the area of high-speed winds blowing west to east

A. humidity

B. jet stream

C. Coriolis effect

D. wind

Match the types of air currents with the descriptions.

1.05) _____ convection current

1.06) _____ surface current

1.07) _____ ocean current

1.08) _____ deep current

1.09) _____ upwelling

A. the movement of nutrients to the surface

B. the movement of the ocean

C. a current that moves deeper than 100 meters in the ocean

D. the movement of water due to difference in density

E. a current that only goes about 100 meters deep into the ocean

Match the types of air masses with the descriptions.

1.010) _____ maritime air mass

1.011) _____ continental air mass

1.012) _____ arctic air mass

1.013) _____ tropical air mass

1.014) _____ low-pressure system

1.015) _____ high-pressure system

1.016) _____ polar air mass

A. an air mass that is not as cold as arctic air and begins over high latitudes over land or sea

B. a cool, dry air system

C. the warm air from low latitudes over land or sea

D. begins over oceans and has a high moisture content

E. the cold air mass from the arctic region

F. a warm, moist air system

G. begins over land and has cold, dry air

Fill in the blanks using words from the box below.

polar jet streams	trade winds	convection	air pressure
atmospheres	precipitation	temperature	sun wind

1.017) ______________________________ blow between the equator and 30° latitudes and curve to the west.

1.018) The energy from the ____________________ drives the convection in the earth's atmosphere.

1.019) The sun provides the energy for ____________________ currents in the ocean.

1.020) ______________________________ bring arctic air down into the United States when they move south.

1.021) The ____________________ is how hot or cold the atmosphere is.

1.022) ____________________ is the movement of air.

1.023) ______________________________ is the weight of the air on the surface of the earth.

1.024) Pressure is measured in ____________________.

1.025) Examples of ____________________ include rain, sleet, snow, and hail.

CHECK ____ CORRECT ____ RECHECK ____

SECTION TWO

Objectives:

- Describe the factors that make up weather.
- Identify how solar energy impacts weather.
- Describe weather systems.
- Identify how weather systems change weather.

Vocabulary:

meteorologist – a scientist who studies weather and the earth's atmosphere

4. GLOBAL WEATHER PATTERNS

Local weather is impacted by global weather patterns. Global weather patterns are widespread conditions that cover thousands of kilometers. These predictable patterns are created by the interactions between the atmosphere and the ocean. One of these interactions is the creation of wind systems. As wind systems move across the earth, they create high- and low-pressure systems that impact global weather patterns. Ocean water temperature and currents also impact global weather patterns. The biggest impacts on global patterns come from global winds, jet streams, and pressure belts. Other impacts come from oceanic events known as El Niño and La Niña.

PRESSURE BELTS

There are four main global pressure belts created by the uneven heating of the earth from the sun (**Fig. 4.1**). These pressure belts influence global winds such as the easterlies and westerlies.

Equatorial Low

At the equator, warm, moist air is rising and forms a low-pressure belt. As that air cools, it forms clouds and precipitation. These pressure belts can generate large amounts of rainfall and form tropical storms.

Subtropical High-Pressure Belt

All deserts are found in this belt since warm, dry weather is common.

Subpolar Low-Pressure Belt

The low-pressure belt is not well-defined all year long. However, the contrast of warm and cold air masses found in this area produces polar jet streams. Polar jet streams greatly influence weather and are discussed further in this Lesson.

Polar High-Pressure Belt

These pressure belts are smaller areas of high pressure found around each pole. They create the cold, dry polar weather.

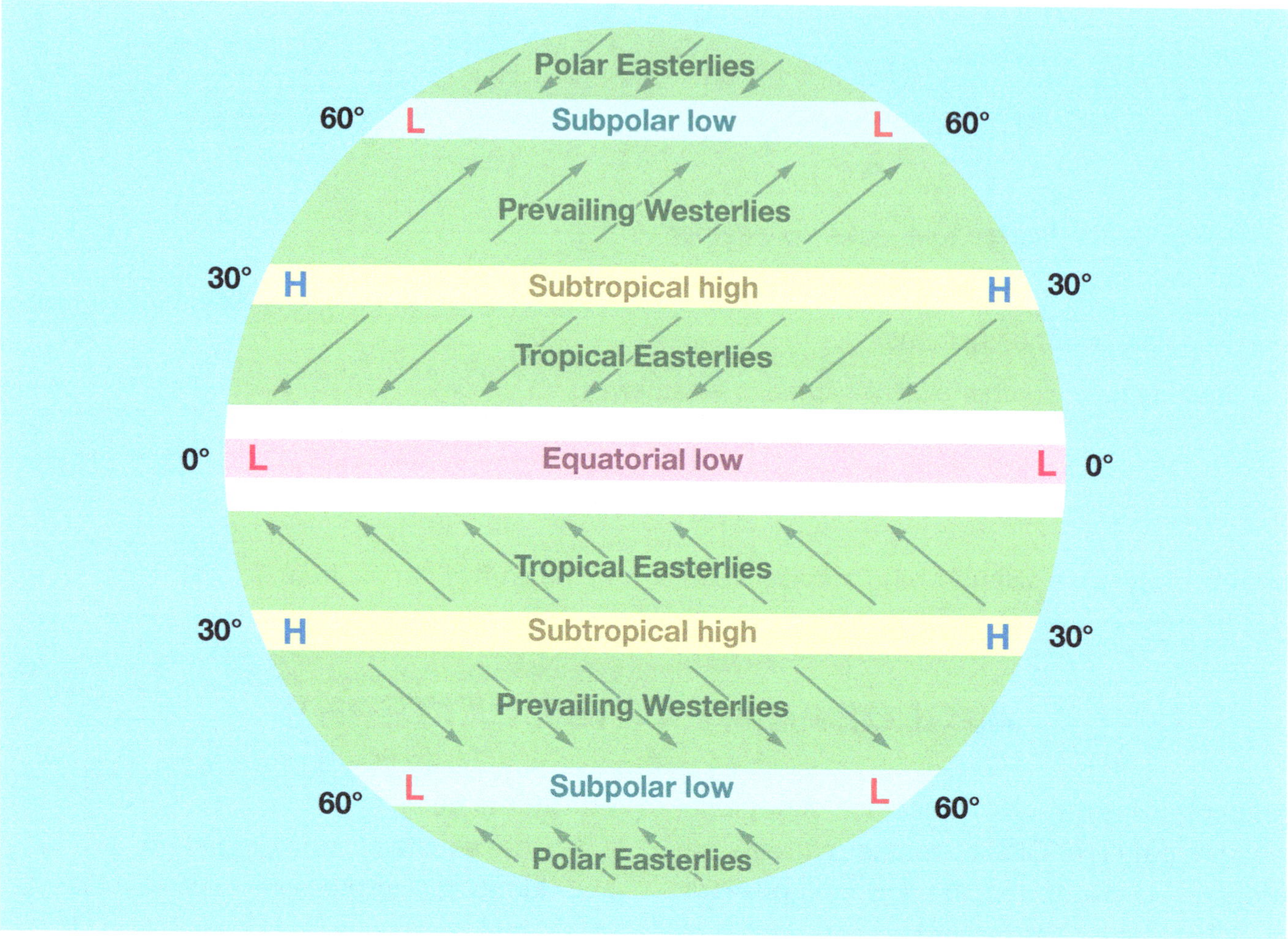

Fig. 4.1 Global pressure belts

Jet Streams

As you learned in Lesson 1, jet streams are areas of high-speed winds blowing west to east (**Fig. 4.2**). There are two different jet streams: the polar jet stream and the subtropical jet stream. These two jet streams are found in both the Northern and Southern Hemispheres. As the seasons change, the position of the jet streams change, staying between the boundaries of hot and cold air. In the Northern Hemisphere, when the polar jet stream moves south, cooler, polar air is also brought south. In the summer, the jet stream moves north, pushing the colder air back up toward the poles. The polar jet stream is responsible for moving weather systems and affects where tornadoes and thunderstorms occur.

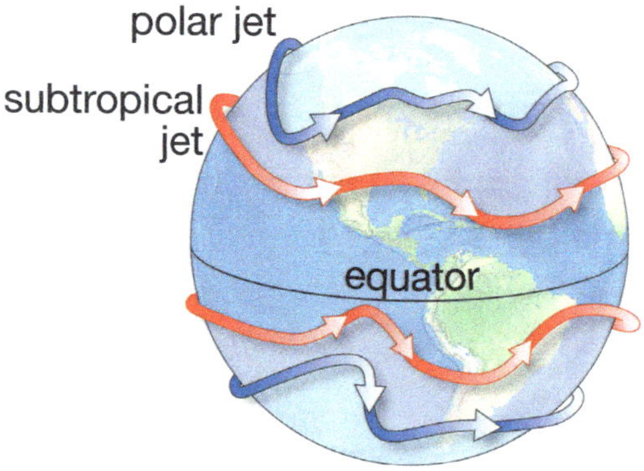

Fig. 4.2 Jet streams

Oceanic Currents

Currents in the ocean also have an impact on the global weather patterns. The convection currents found in the ocean move heat energy from place to place, affecting the temperature and weather in coastal areas. Surface currents move water horizontally across the ocean.

As surface winds move, they determine the flow of the surface currents. Warm surface currents move warm water from the low latitudes to the high latitudes.

The Gulf Stream is a warm-water current that impacts weather on the United States' East Coast and Northwestern Europe (**Fig. 4.3**). Winds from this current blow from east to west and keep the Florida winter temperatures mild. The Gulf Stream also creates tropical storms as energy is transferred from the ocean to the atmosphere. Solar energy allows water at the ocean's surface to evaporate and be released as water vapor and becomes the driving force for the development of storms.

The California Current is a cold-water current that travels along the west coast of the United States (**Fig. 4.4**). Winds in this current move the surface water away from the shore, allowing deep cold water to rise to the surface and reduce the temperature of the California Current water even more. The cool water temperatures make the summer temperatures along the United States' west coast cooler than inland

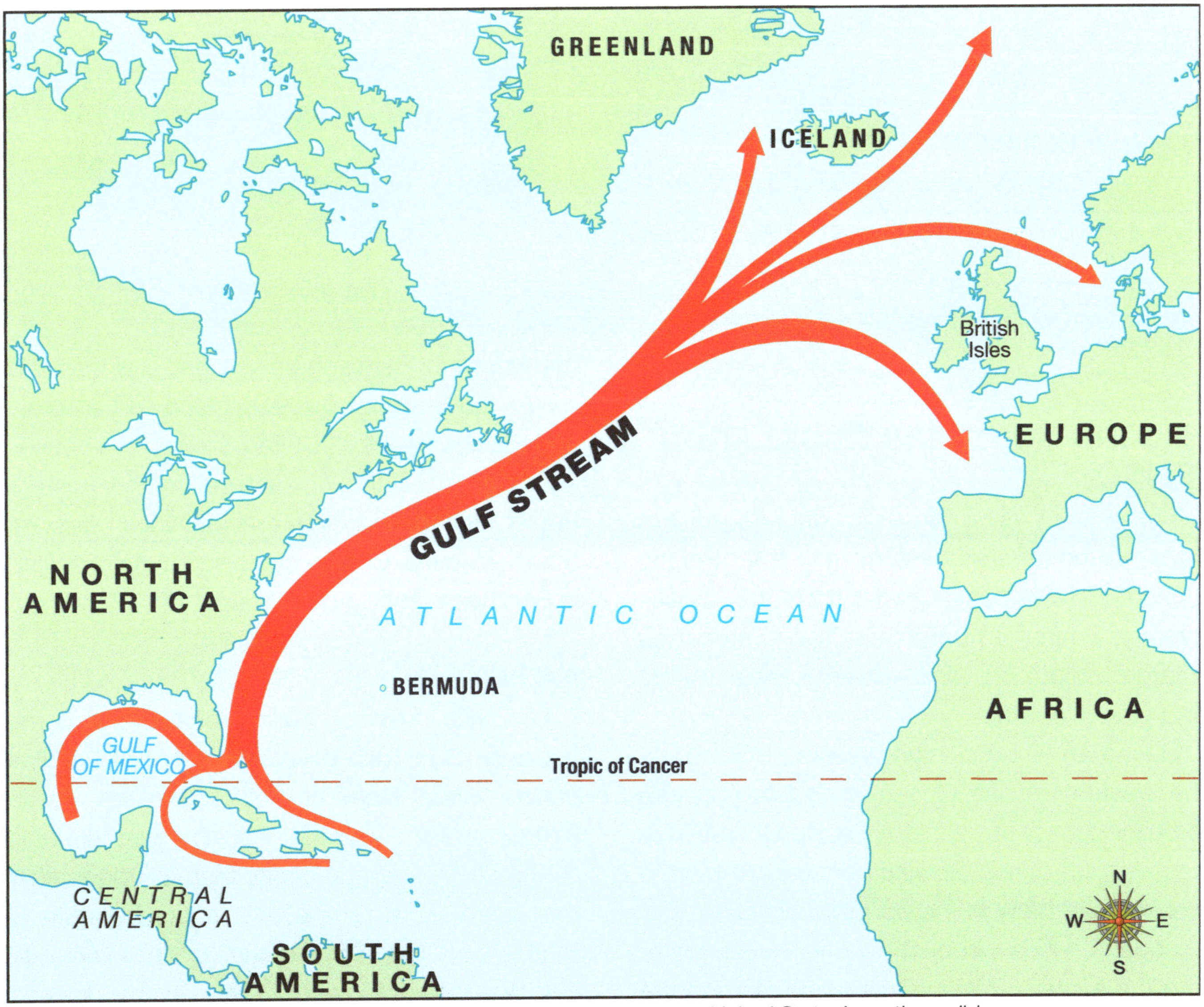

Fig. 4.3 The Gulf Stream helps keep the eastern United States' weather mild.

Source: Mountain High Maps® Copyright © 1993-2020 Digital Wisdom®, Inc.

areas. Due to the cold temperatures of the California Current, tropical storms do not have the energy to form. Storms that approach the coast weaken but bring rain to the coastline.

Fig. 4.4 The California Current carries cold Pacific Ocean water south toward Mexico. This keeps the temperature in California mild and not too hot.

Source: Mountain High Maps® Copyright © 1993-2020 Digital Wisdom®, Inc.

El Niño Southern Oscillation (ENSO)

The ENSO is a periodic change in the ocean-atmosphere system in the Pacific Ocean. El Niño occurs when there are above-average ocean temperatures. This system has a large impact on wind patterns, jet streams, and global weather.

During an El Niño, high pressure builds over the western Pacific Ocean while low pressure is building over the eastern Pacific Ocean. This causes trade winds blowing from east to west to weaken and blow in the opposite direction. The change in winds will push warmer ocean water into the eastern Pacific Ocean and change the pattern in the Pacific Ocean and change the pattern in the jet stream. The polar jet stream in the Northern Hemisphere will get stronger and move south, bringing precipitation to California and the southeastern United States. This causes flooding, more tropical storms in the eastern Pacific Ocean, and more precipitation in western South America. El Niño can also cause droughts in Southeast Asia and Australia.

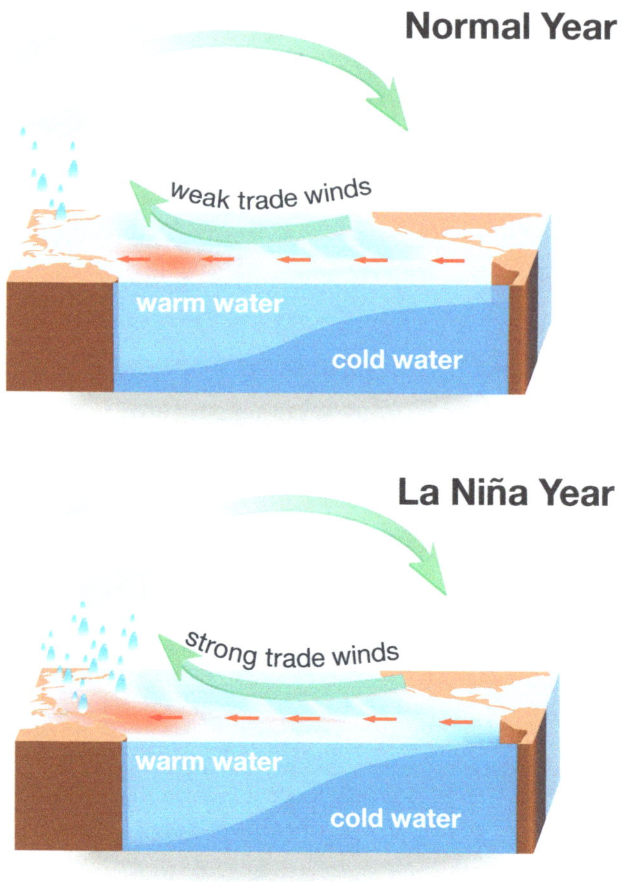

Fig. 4.5 La Niña

La Niña

La Niña events have weather patterns opposite of El Niño events (**Fig. 4.5**). A La Niña occurs when there are below-average ocean temperatures. Tropical Pacific temperatures during La Niña will be cooler than average rather than warmer. La Niña events occur between El Niño events. The effects of the La Niña events are also opposite of those of El Niño events. Instead of floods, droughts occur in areas that would normally receive more rainfall during an El Niño.

Likewise, floods often occur in areas that usually experience droughts during El Niño. However, while La Niña events seem to be exact opposites to El Niño events, they are never as strong as El Niño events and are harder to predict. La Niña events also do not cause the same amount of impact and consequences that El Niño events do.

REVIEW

Choose the best answers.

4.1) _____ There are four different ___ that form between latitudes.
A. jet streams
B. currents
C. Coriolis effects
D. pressure belts

4.2) _____ The warm water current that moves from the United States' East Coast to Northwestern Europe is the ___.
A. Gulf Stream
B. California Current
C. La Niña
D. El Niño

4.3) _____ The ___ travels along the West Coast and brings cold water temperatures.
A. Gulf Stream
B. California Current
C. La Niña
D. El Niño

4.4) _____ ___ occurs when there are above-average ocean temperatures.
A. Gulf Stream
B. California Current
C. La Niña
D. El Niño

4.5) _____ ___ occurs when there are below-average ocean temperatures.
A. Gulf Stream
B. California Current
C. La Niña
D. El Niño

Match the descriptions with the words.

4.6) _____ brings cold air to the northern United States
4.7) _____ gives Florida mild winters
4.8) _____ does not allow tropical storms to form
4.9) _____ causes droughts in South America
4.10) _____ causes more tropical storms in eastern Pacific Ocean

A. California Current
B. El Niño
C. Gulf Stream
D. La Niña
E. polar jet stream

CHECK ☐ CORRECT ☐ RECHECK ☐

5. LOCAL WEATHER

Have you ever had a soccer or football game get canceled because of a storm? Weather, especially local weather, impacts our daily lives. Temperature, humidity, precipitation, air pressure, and wind all contribute to the weather that impacts where we live.

Because weather across the globe is interconnected, oftentimes weather events happening thousands of miles away can affect local weather patterns. The term "local weather" is used to describe the current atmospheric conditions in our immediate area at a given point in time, such as a small city or town.

The daily weather conditions that we experience in our local area are the result of a combination of many different factors. Major factors of weather occur on a variety of size and time scales, including:

- changes in climate that take place over many years, decades, and centuries
- seasonal changes that are brought on by the earth's position in relation to the sun, lasting for several months
- large-scale global weather patterns that occur over days, weeks, and months
- local and regional effects, which include daytime heating, geography, vegetation, and urbanization

We already learned how large-scale, global weather patterns can provide clues as to what type of weather we might expect to receive in a given season. In this Lesson, we will look at how global weather patterns influence local weather and how clouds, precipitation, wind, and thunderstorms are types of local weather that can impact your day.

Global Winds

Almost all weather is the result of differences in temperatures (heating) between regions of the earth, both locally and globally. The spherical shape and tilt of the earth create differences in how the sun hits it. This uneven heating of the planet produces global circulation patterns that cause the continuous movement of air around the world.

Local weather is largely influenced by global patterns of atmospheric movement such as jet streams, polar easterlies, westerlies, and trade winds. Atmospheric jet streams, for example, can help transport different air masses into various regions. When these winds move air masses, they carry their weather conditions (hot or cold, dry or moist) from one location to another. The most active weather occurs where these air masses interact or collide with each other and may create powerful storms.

Just like oceans and rivers, jet streams are fast-moving currents high in the atmosphere. Jet streams carry weather systems, moving heat and moisture all across the globe.

These fast-flowing, narrow bands of wind are largely responsible for the day-to-day changes in weather. They affect the weather by impacting temperatures and precipitation. Because it is generally colder and wetter north of a jet, southward-moving jets help to push cold air south. Northward-moving jets help push warm air north, as it is typically drier and warmer south of the jet.

Monitoring jet streams can help meteorologists, or people who predict weather, determine where weather systems will move next. The path of a jet stream can change quickly, taking storms in different directions.

CLOUDS

The presence of clouds indicates that a weather system is in place. There are many different types of clouds that signal different types of weather. Wispy cirrus clouds signal mild weather, while nimbostratus clouds create long periods of precipitation. Clouds can block the amount of sunlight that reaches the earth's surface, making cloudy days cooler.

There are three main types of clouds:

1 – Cumulus clouds are puffy, like cotton balls. A blue sky with small, white cumulus clouds can mean the weather will be fair and warm with no rain.

2 – Stratus clouds form in large layers, like sheets. When the whole sky is gray, it is usually covered by a type of stratus cloud. Sometimes, rain can fall from stratus clouds, but it will not be very heavy.

3 – Cirrus clouds look like wisps of hair. They are thin, long, and found high in the atmosphere. Cirrus clouds appear on sunny but slightly cooler days.

Some clouds are combinations of different types of clouds and are classified by where they occur in the atmosphere. There are two prefixes that can be found in front of a cloud's name:

- *alto* – mid
- *nimbo* – rain

A nimbostratus (*nimbo*– + *strato*) cloud would mean the cloud looks like a layer and will bring rain. An altocumulus (*alto*– + *cumulo*) would be in the middle of the sky (not high or low) and look puffy. Large cumulonimbus clouds, also called thunderheads, create heavy rains, thunderstorms, and tornadoes.

REVIEW

Choose the best answers.

5.1) _____ What type of cloud does the image depict?

A. stratus
B. cirrus
C. cumulus

5.2) _____ What type of cloud does the image depict?

A. stratus
B. cirrus
C. cumulus

5.3) _____ What type of cloud does the image depict?

A. stratus
B. cirrus
C. cumulus

5.4) Which THREE types of clouds can produce rain?

- ☐ cirrus
- ☐ cumulus
- ☐ stratus
- ☐ cumulonimbus
- ☐ nimbostratus

5.5) _____ Storm systems that move across the United States are primarily driven by ___.

A. ocean currents
B. jet streams
C. cumulus clouds
D. local geography

PRECIPITATION

The more condensed water a cloud has, the darker the cloud becomes. When enough water has condensed in a cloud, the cloud can no longer "hold" the water in. The water then falls as precipitation. Depending on temperature, precipitation can fall as snow, freezing rain, hail, or sleet.

There are two ways in which water in a cloud can become rain. In parts of the earth where the air is too warm to condense water into ice crystals, the water molecules attract other molecules and stick together. Like a snowball getting bigger as it rolls down the side of a snowy mountain, a water drop gets bigger as it runs into other water drops. When the collector drop (the original drop) collects enough water, it eventually becomes too heavy and falls from the cloud.

> As you have seen throughout this Unit, a common theme with weather is the sun. All weather begins with the sun. The sun brings energy to the earth in the form of heat and light. The sun does more than provide sunny days, however. Winds, ocean currents, storms, and hurricanes all begin with the uneven heating of the earth by the sun. This creates an imbalance, which always seeks to balance itself through convection currents. Hot air runs into the cold air, causing wind. Warmer water will flow to colder water, causing currents.

Clouds can also become too heavy with another process called the Bergeron process (**Fig. 5.1**). If water vapor forms high enough in the atmosphere where temperatures are colder, the vapor will condense into ice crystals. These ice crystals attract more water vapor to them and get bigger. The ice crystals continue to grow and eventually fall from the clouds as precipitation. As the ice crystals fall, the air temperature gets warmer and the ice crystals melt into rain. If the air temperature stays below freezing all the way to the ground, then sleet or snow will fall.

Fig. 5.1 Most precipitation is made by the Bergeron process.

THUNDERSTORMS

Thunderstorms are characterized by strong winds, lightning and thunder, and sometimes tornadoes. In order for a thunderstorm to form, the air must be unstable and have moisture. Increased moisture in the air occurs after especially hot days, which causes increased ocean water evaporation. There are different types of thunderstorms, but only the main types will be discussed in this Lesson.

Single-Cell Thunderstorms

These are small storms that only last a short time. They often occur from heat and evaporation on a hot summer afternoon. These types of thunderstorms may have a brief time of heavy rain and lightning.

Most single-cell storms are not severe, but some can be. In order for a storm to be severe, it must produce one or more of the following: hail that is at least one inch wide, winds that are at least 58 miles per hour, or a tornado.

Multicell Storms

Most thunderstorms fall under this category. Multicell storms are a group of single-cell storms that move together as one unit. Multicell storms can make medium-size hail, cause flash floods with torrential amounts of rainfall, or produce weak tornadoes. Like single-cell storms, multicell storms can be severe or mild.

Squall Line

A squall line is a special type of multicell storm (**Fig. 5.2**). This is a group of thunderstorms with a wide and well-developed front and strong winds called a gust front. Like multicell storms, squall lines produce hail, weak tornadoes, and sometimes cause flash flooding.

Supercell Storms

To be a supercell storm, there must be some rotation in the updraft. Supercell storms are always severe thunderstorms. Strong tornadoes and powerful winds come out of supercell storms. Supercell storms are often long lasting (more than one hour).

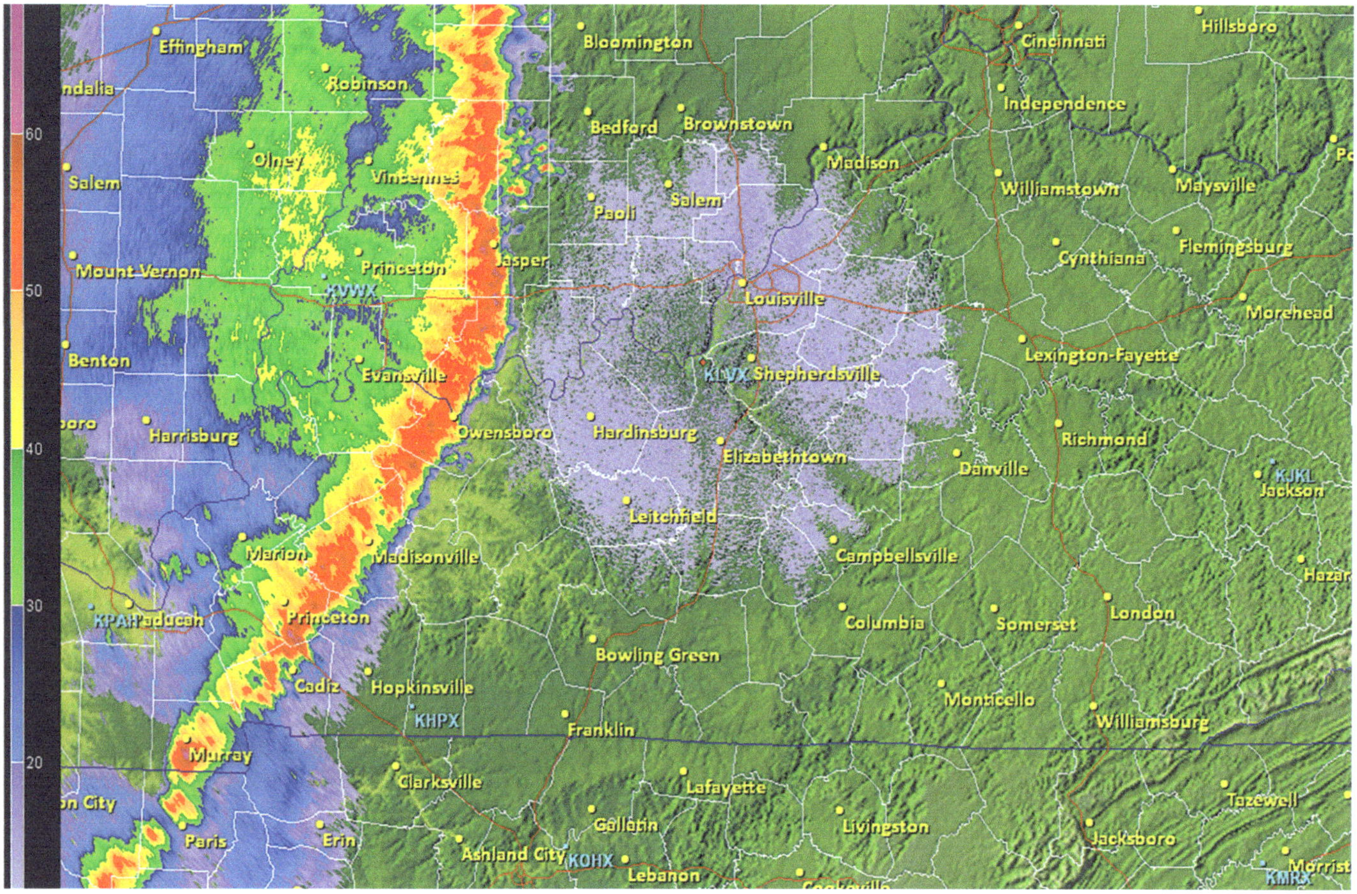

Fig. 5.2 A squall-line thunderstorm

Choose the best answers.

5.6) _____ All weather begins with the ___.

A. sun
B. clouds
C. ocean
D. precipitation

5.7) _____ Winds, ocean currents, storms, and hurricanes all begin with the ___ heating of the earth by the sun.

A. magnetic
B. uneven
C. even
D. gravitational

5.8) _____ In order for a thunderstorm to form, the air must ___.

A. have moisture
B. be dry
C. be unstable
D. both A and B
E. both A and C

5.9) _____ What type of thunderstorm produces strong tornadoes?

A. single-cell
B. multicell
C. squall line
D. supercell

5.10) _____ Which way is most rain formed?

A. Condensed water droplets collect other water droplets to form the collector drop until they fall from the clouds as rain.
B. Condensed ice crystals collect other ice crystals until they fall from the clouds and then melt into rain when they hit a layer of warm air.

5.11) _____ Which way is rain formed in the tropics where the air is warmer?

A. Condensed water droplets collect other water droplets to form the collector drop until they fall from the clouds as rain.
B. Condensed ice crystals collect other ice crystals until they fall from the clouds and then melt into rain when they hit a layer of warm air.

5.12) Which THREE qualities make a thunderstorm severe?

- ☐ any hail
- ☐ hail at least one inch wide
- ☐ any tornado
- ☐ winds that are at least 58 miles per hour
- ☐ a squall line

CHECK ☐ CORRECT ☐ RECHECK ☐

Character Spotlight

What does determination look like in practice? It is actually fairly complicated. First, it requires a desire to do something. Second, it means deciding whether the thing you want to do is a right and proper thing, and whether it is worth an all-out effort. Third, it requires a firm resolve to do it. Fourth, it requires an evaluation of all the obstacles that might arise. Fifth, it means figuring out what skills you do not have that you will need in order to reach your goal. Finally, it means you must learn to work. You must decide that what you are going to do is worth doing and that you will not let anything stop you. Then, go for it! As Davy Crockett said, "Be sure you are right—then go ahead."

6. METEOROLOGY AND MAPS

When watching the news, a meteorologist stands in front of maps and talks about the weather for the day and the week. A **meteorologist** is a scientist who studies weather and the earth's atmosphere. Sometimes, a meteorologist's predictions are accurate. However, meteorologists can get weather forecasts wrong because forecasts are predictions, not statements of facts. These predictions are made by analyzing data and looking for weather patterns and systems. Then, using prior knowledge, they can make a forecast or prediction about what will happen.

The data a meteorologist uses includes: temperature, humidity, wind direction, wind speed, clouds, precipitation, and atmospheric pressure. By observing these elements, a weather forecaster is able to predict the changes in our weather needed for people to be safe and comfortable as they go through the week.

A meteorologist is a scientist who studies weather and the earth's atmosphere.

HOW IS WEATHER DATA COLLECTED?

In order to make accurate predictions, meteorologists need accurate data from different sources. Meteorologists use different types of technology to gather data at various locations. Ground stations are called automated surface stations that collect weather data from the lower atmosphere layers. A ground station will include instruments that measure pressure, temperature, humidity, precipitation, wind speed, visibility, and cloud cover.

A ground station

Radar is used to monitor the location, movement, and intensity of storm systems. Weather radars bounce radio waves off the precipitation inside of a storm (**Fig. 6.1**). The more precipitation in the storm, the stronger the signal will be recognized by the radar. Doppler radar is a type of weather radar that can detect precipitation and air motion in a storm. This radar can also be used to monitor severe storms and tornadoes.

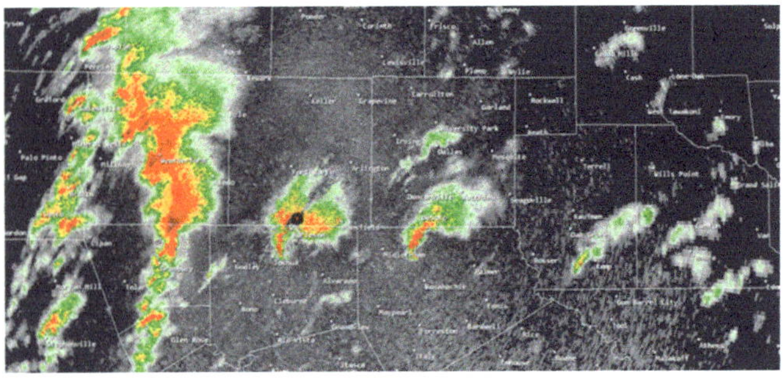

Fig. 6.1 This radar map shows areas of precipitation and the intensity (relative amount) that is currently happening over an area.

To reach the middle and upper atmospheres, balloons and airplanes carry weather-sensing instruments to measure weather conditions. Weather balloons are sent into the air twice daily from weather stations. These balloons collect information at different altitudes with radiosonde instrument packages. Radiosondes measure the atmospheric pressure, air temperature, humidity, wind speed, and direction up to about 32 kilometers in the atmosphere. They then send the data by radio signal to ground stations.

Weather satellites orbit at high altitudes and provide data on water vapor, temperatures from the top of the clouds, and the movement of weather systems. There are two satellites monitoring the weather: geostationary satellites and polar orbiting satellites. Geostationary weather satellites monitor from a fixed position above the earth, while polar orbiting satellites circle the earth to provide global data. Images used to track and report on weather conditions are taken from cameras found on the satellites.

A weather balloon

A weather satellite

REVIEW

Choose the best answers.

6.1) _____ A scientist who studies weather and the earth's atmosphere is called a(n) ___.

A. atmospherologist
B. weatherologist
C. meteorologist
D. climatologist

6.2) _____ ___ are automated surface stations that collect weather data from the lower atmosphere layers.

A. Radar stations
B. Observatory stations
C. Weather stations
D. Ground stations

6.3) _____ A(n) ___ is used to monitor the location, movement, and intensity of storm systems.

A. upper air chart
B. station model
C. weather balloon
D. radar

6.4) _____ A weather radar that detects precipitation and air motion is called a(n) ___.

A. air chart
B. Doppler radar
C. weather balloon
D. red shift

6.5) _____ ___ are sent up twice daily from weather stations to collect data.

A. Air charts
B. Doppler radars
C. Weather balloons
D. Red shifts

Weather Maps

Weather maps are used to display weather data and make it easier to understand. The symbols found on weather maps are called station models; they represent the weather found at a particular weather station. On a weather map, a station is represented by a small circle and is surrounded by symbols and numbers that describe the current weather data at that location. The main elements found on a station model include temperature, wind speed, wind direction, cloud cover, air pressure, and dew point.

Surface weather maps are the maps you see on your local news (**Fig. 6.2**). Surface weather maps display air pressure, precipitation, and the location of fronts. Air pressure is represented with isobars, which are lines with connected points of equal air pressure. Isobars form closed loops with the center of the loop having an uppercase "H" for high pressure or an uppercase "L" for low pressure. Recall that clear and sunny weather occurs in high-pressure systems, and cloudy or stormy weather occurs in low-pressure systems.

Cold fronts are represented with blue lines that have blue triangles. Warm fronts are represented with red lines and half circles. Stationary fronts alternate between red and blue color lines, with both triangles and half circles.

Upper air charts are weather maps that display data collected from instruments on weather balloons. Upper air charts show wind and air pressure from the middle and upper levels of the atmosphere and help determine where weather systems will form, move, or fall apart. These charts can also be used to identify the jet stream position, and they are used by airline pilots to identify flight paths and air turbulence (rough, bumpy air).

Weather Forecasts

Today, meteorologists have access to supercomputers that have increased the accuracy of weather predictions. Meteorologists and weather agencies use these computers to form different types of forecasts. Short-range forecasts make weather predictions for the present day and up to three days out. Medium-range forecasts are for three to seven days out, while long-range forecasts are for weeks and months into the future. When making a weather forecast, meteorologists predict the temperature, wind, cloud cover, and precipitation.

Forecasting is an imperfect science. There are many variables that impact the weather and can cause it to constantly change. Due to these changing conditions, short-term forecasts are more accurate than long-term forecasts.

Severe weather forecasts are an important aspect of a meteorologist's job. People need to be warned of any dangerous weather approaching, such as severe storms, tornadoes, and hurricanes, in order to make preparations. In the event of severe weather, The National Weather Service (a federal government agency) will issue weather advisories, watches, or warnings.

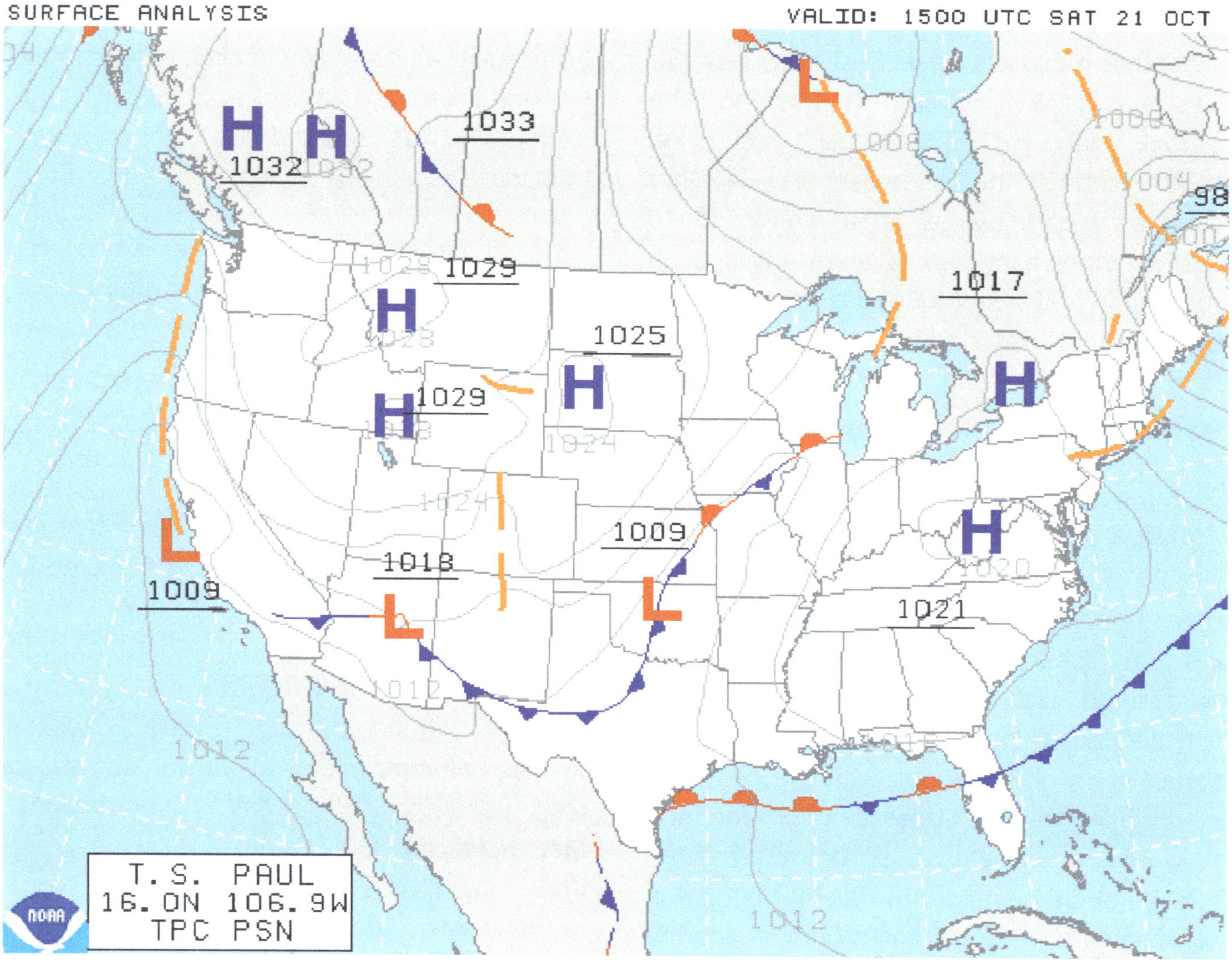

Fig. 6.2 Surface weather map

- A weather advisory is issued for weather conditions that are not serious but can cause an inconvenience, such as a heat advisory.
- A weather watch is issued when severe weather conditions are possible over a large area, such as a thunderstorm watch in the spring.
- A weather warning is issued when the weather conditions are occurring and pose a threat to life and property. This means that people in the line of the storm need to take immediate action, such as with a tornado warning.

Write the correct answers using the map.

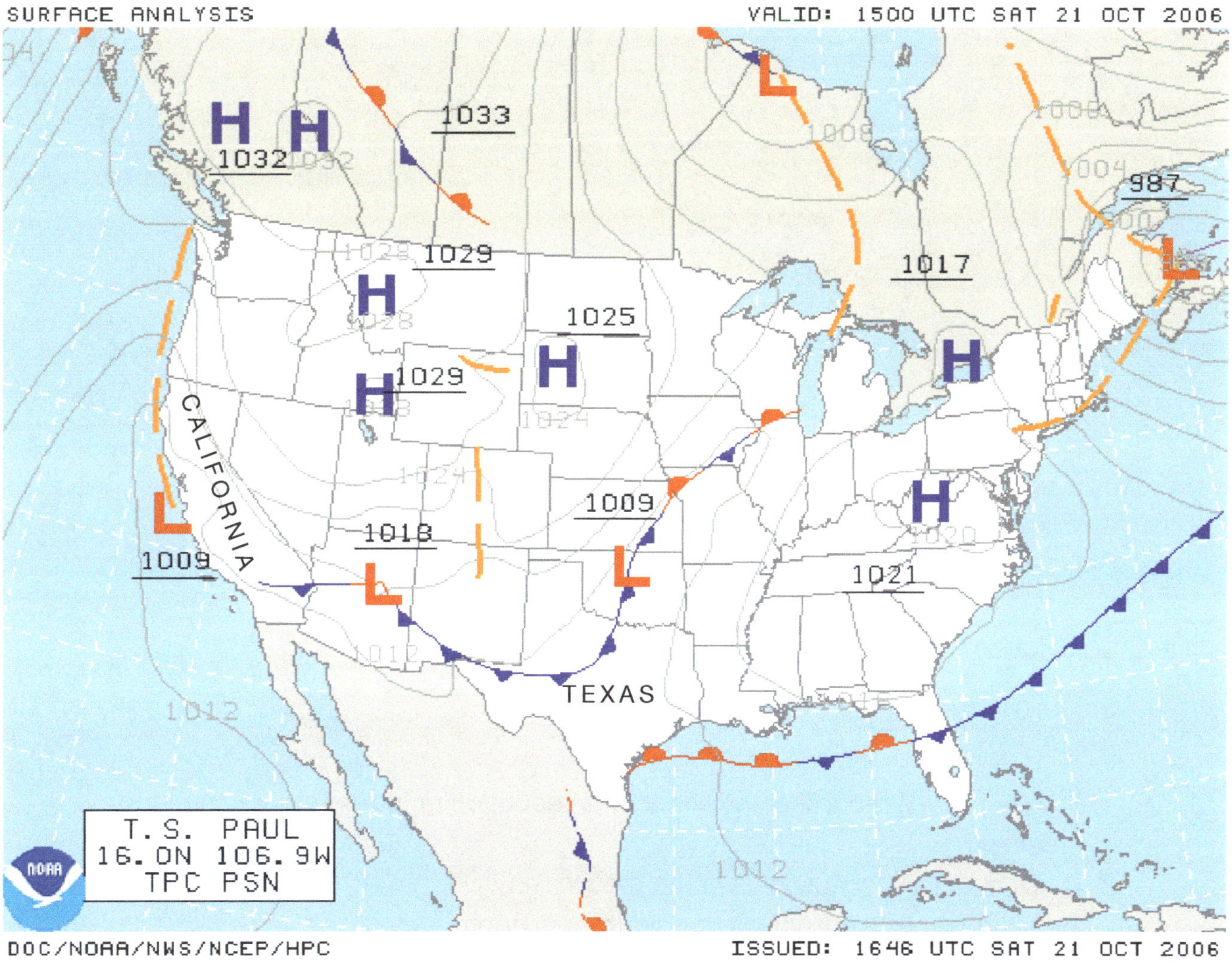

6.6) What type of front is moving through Texas? ____________________

6.7) Where in the United States are the stationary fronts? ____________________

6.8) What type of pressure system is on the coast of California? ____________________

6.9) What type of pressure system is found in the northwest? ______________________________

____

Match the statements with the types of severe weather forecast.

6.10) _____ Temperatures will hit 105 °F at 4:00 p.m.

6.11) _____ A thunderstorm could develop this afternoon.

6.12) _____ A tornado has been spotted outside of town.

A. advisory

B. warning

C. watch

Choose the best answers using the map.

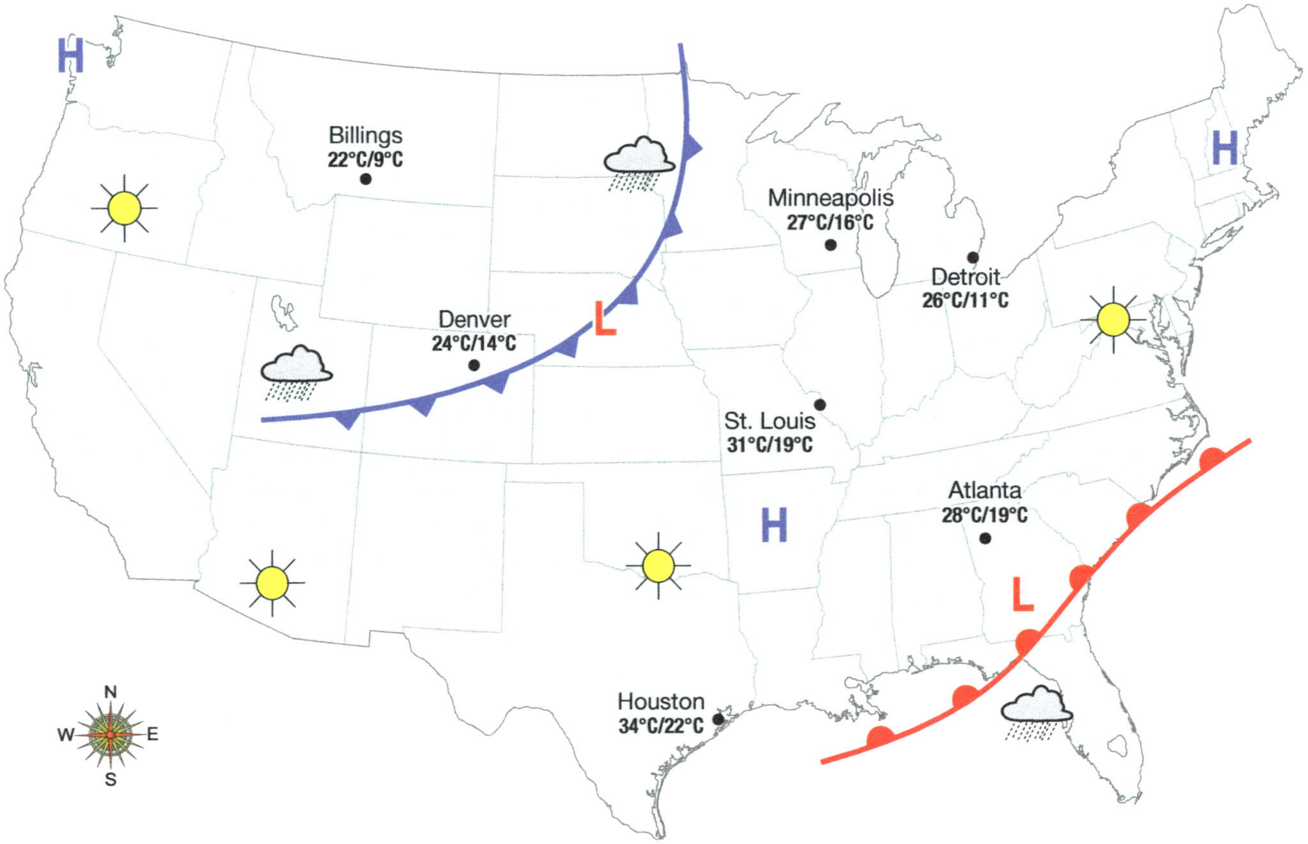

Source: Mountain High Maps® Copyright © 1993-2020 Digital Wisdom®, Inc.

6.13) _____ Which city will most likely experience a reduction in temperatures and stormy weather?

A. Denver will because it is in the mountains.

B. Minneapolis will because a cold front is approaching.

C. Atlanta will because a warm front just went through.

D. Houston will because it is receiving moisture from the Gulf of Mexico.

6.14) _____ Why is Texas experiencing clear and sunny weather?

A. There is a low-pressure system in place.

B. A warm front just moved through.

C. There is a high-pressure system in place nearby.

D. It is receiving moisture from the Gulf of Mexico.

CHECK CORRECT RECHECK

Character Spotlight

Winston Churchill was the inspirational leader of the people of England during World War II. He made tremendous speeches to bolster the spirit of determination in his people. The finest example of these speeches is his speech called "Blood, Toil, Tears, and Sweat."

"I have nothing to offer but blood, toil, tears and sweat. We have before us an ordeal of the most grievous kind. We have before us many, many long months of struggle and of suffering. You ask, what is our policy? I can say: It is to wage war, by sea, land and air, with all our might and with all the strength that God can give us; to wage war against a monstrous tyranny, never surpassed in the dark, lamentable catalogue of human crime. That is our policy. You ask, what is our aim? I can answer in one word: It is victory, victory at all costs, victory in spite of all terror, victory, however long and hard the road may be."

ACTIVITY: Weather Symbols Map

Use the image and information provided to construct a map of the weather forecast across the country for today.

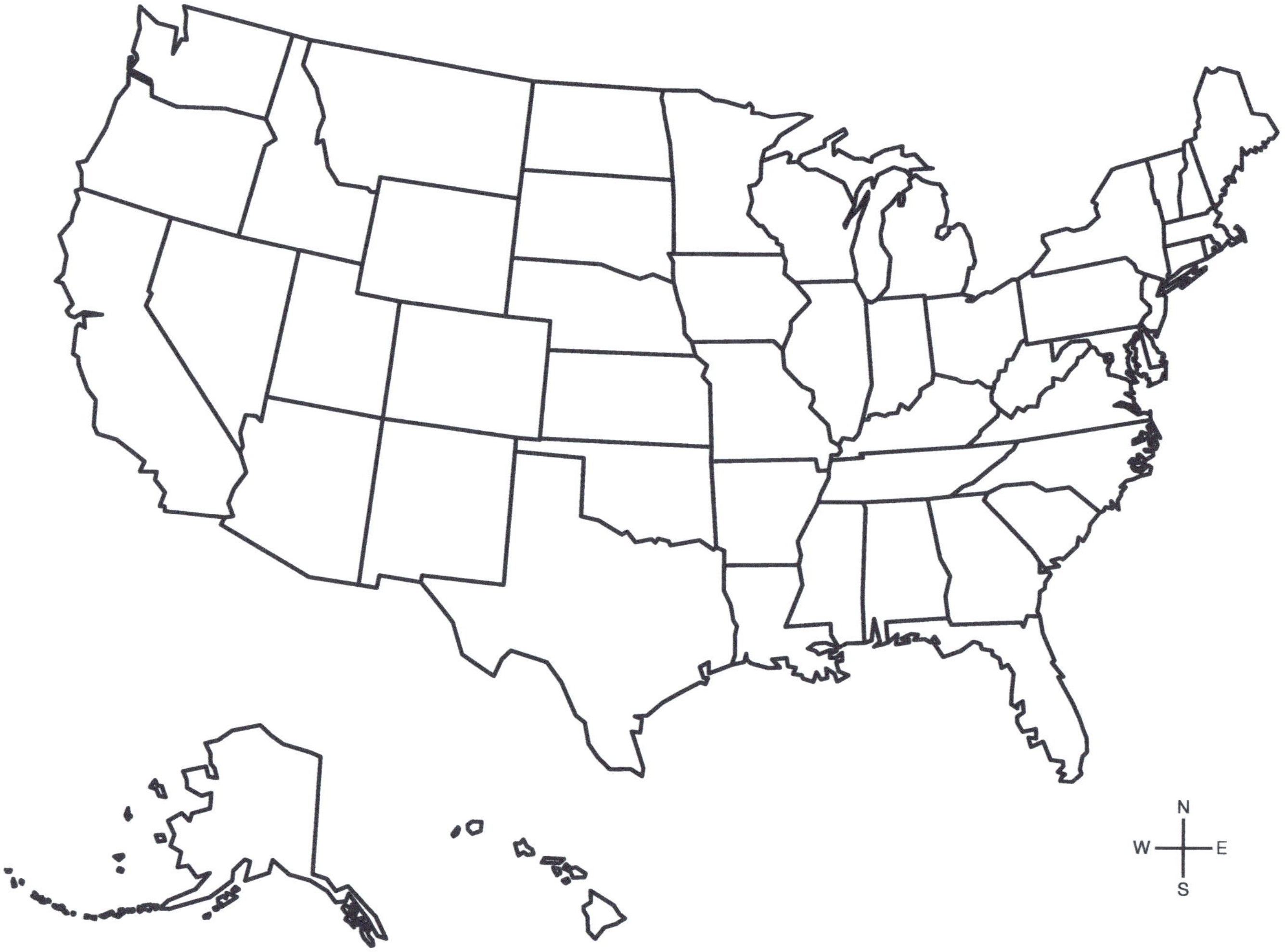

Materials:

- colored pencils

Directions:

1 – Draw showers and thunderstorms along the East Coast from Maine to Virginia.

2 – Draw a low-pressure system over the Pacific Ocean moving into Southern California.

3 – Draw a cold front moving through Texas.

4 – Draw clear, blue skies over Florida with temperatures in the 80s.

5 – Draw a high-pressure system over South Dakota.

6 – Draw a warm front moving from Washington state toward Wyoming.

7 – Draw a stationary front over Louisiana.

8 – Create a key for your weather map on the table provided.

9 – Answer the conclusion questions that follow.

WEATHER SYMBOLS KEY	
Cold Front	
Warm Front	
Stationary Front	
High Pressure	
Low Pressure	

Conclusion:

Write the correct answers.

1) What type of weather would you expect from a cold front? ______________________________

____

____

2) What type of weather would you expect from a warm front? ______________________________

____

____

3) What type of weather would you expect from a stationary front?____________________________

____

____

4) What type of weather would a high-pressure system bring? ______________________________

____

____

5) What type of weather would a low-pressure system bring? ______________________________

____

____

6) How are weather maps useful? ____

____

____

☐ TEACHER CHECK

CHARACTER SPOTLIGHT

As an eighth-grader, Polly decided that she was going to graduate from high school at the top of her class. She wanted to be a doctor, and in order to get into medical school, she would need high grades. To ensure her success, Polly asked her father to hire a math tutor for her, even though she was a fairly good math student.

The added instruction proved helpful, allowing her to get top grades in math and science through four years of advanced classes in high school. Polly also participated in extra-curricular activities during her years in school, including serving on the student council, but her determination never wavered. She was the school's valedictorian and gave a speech at her commencement exercises—all of this despite the fact that Dr. Polly Myers was born totally deaf.

MY SCORE

QUIZ 2

TEACHER INITIAL

(Each answer, 5 points)

Fill in the blanks using words from the box below.

collector drops	ice crystals	sun	moisture
cumulonimbus	thunderstorms	hail	cumulus

2.01) All weather begins with the ______________________.

2.02) Most rain is formed from ________________________________ that melt as they fall closer to the surface.

2.03) Rain can fall from ______________________ clouds.

2.04) Rain in warmer, subtropical areas forms from _______________________________.

2.05) Supercells are severe ______________________ that can produce tornadoes.

2.06) The air must be unstable and have ______________________ for thunderstorms to form.

Match the words with the descriptions.

2.07) _____ cold front

2.08) _____ weather balloon

2.09) _____ radar

2.010) _____ meteorologist

A. sent up twice daily from weather stations to collect data

B. studies weather and the earth's atmosphere to predict weather forecasts

C. causes temperatures to fall and thunderstorms to develop

D. monitors the location, movement, and intensity of storm systems

Match the words relating to global weather patterns with the descriptions.

2.011) _____ El Niño

2.012) _____ California Current

2.013) _____ Gulf Stream

2.014) _____ La Niña

2.015) _____ pressure belt

A. forms between latitudes

B. allows Florida to have mild winter temperatures

C. occurs when ocean temperatures are below average

D. occurs when ocean temperatures are above average

E. stops the formation of tropical storms by bringing cold water temperatures

Match the weather-related words with the descriptions.

2.016) _____ ground station

2.017) _____ Doppler radar

A. detects precipitation and air motion

B. automated surface station that collects weather data from the lower atmosphere layers

Choose the best answer.

2.018) _____ Areas of high-speed winds blowing west to east that influence weather patterns are called ___.

A. polar easterlies
B. prevailing westerlies
C. jet streams
D. trade winds

Choose the best answers using the map.

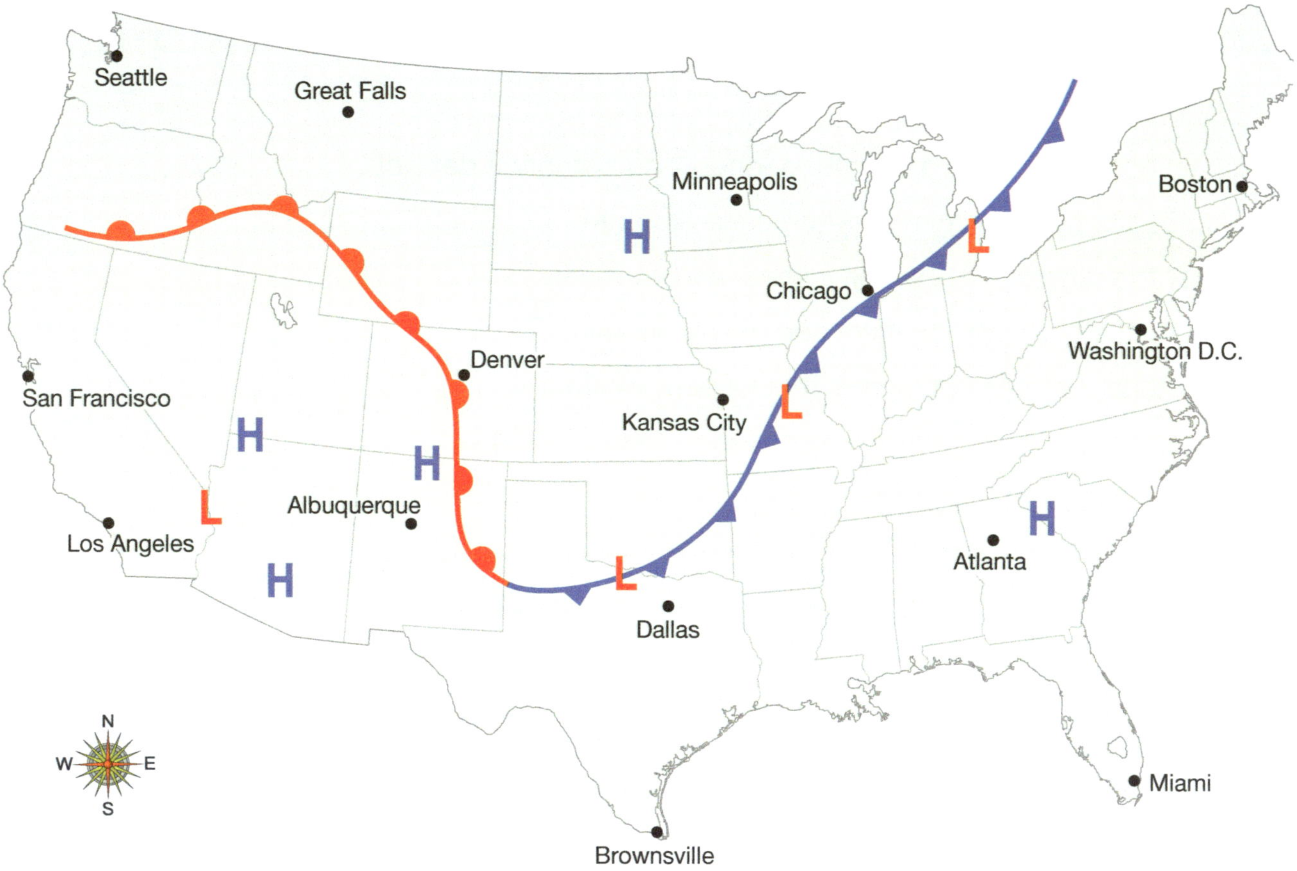

Source: Mountain High Maps® Copyright © 1993-2020 Digital Wisdom®, Inc.

2.019) _____ Which city is experiencing a warm front?

A. Denver
B. Minneapolis
C. Chicago

2.020) _____ Which city is likely experiencing stormy weather?

A. Denver
B. Minneapolis
C. Chicago

CHECK ____ CORRECT ____ RECHECK ____

SECTION THREE

Objectives:

- Explain how hurricanes, tropical cyclones, and typhoons develop.
- Describe the scale system and hurricane damage.
- Identify the difference between climate and weather.
- Describe the greenhouse effect and global warming.
- Understand the different factors that influence climate.
- Understand the main factors that influence climate change.
- List the evidence for a warmer earth.
- Describe both natural and human factors for climate change.
- Understand possible future consequences of climate change.

Vocabulary:

anthropogenic *[an-thruh-puh-JEN-ik]* – an event or change that is caused by humans

climate – the weather conditions in an area over a long period of time

climate change – a long-term change in a region's precipitation and temperature

global warming – the rising of the overall global temperature

greenhouse effect – the trapping of heat that reradiates from the surface of the earth

hurricane – a low-pressure storm system in the Atlantic Ocean and North Pacific Ocean that produces winds of 74 miles per hour or more

storm surge – a large wave created by a hurricane

tipping point – the point at which a series of small changes causes a larger, more important change

7. HURRICANES, TROPICAL CYCLONES, AND TYPHOONS

In 2017, Hurricane Harvey made landfall in southern Texas. The storm became the second costliest hurricane to hit the United States. It dumped over 27 million gallons of rain in Texas. The National Weather Service had to make a new color category to show the over 50 inches of rain that fell in some parts of Houston. It rained so much that the city of Houston temporarily sank almost an inch under all the weight of the water. The storm created about 8 million cubic yards of trash in Houston from all the items destroyed by flooding.

Hurricane Harvey was a tropical cyclone. Tropical cyclones are low-pressure systems that form over warm, tropical oceans. Both hurricanes and typhoons are tropical cyclones. The difference between a hurricane

In the United States, we refer to tropical cyclones as hurricanes. However, tropical cyclones have different names depending on where the storm formed (**Fig. 7.1**).

- In the Atlantic Ocean, all tropical cyclones are called hurricanes.
- In the Indian Ocean, tropical cyclones are just called tropical cyclones.
- In the Pacific Ocean, tropical cyclones have different names based on where they form in the Pacific.
- Cyclones that form off the West Coast of the United States (eastern Pacific Ocean) are called hurricanes.
- Cyclones that form in the north/northwest Pacific Ocean are called typhoons.
- Cyclones that form in the south/southwest Pacific Ocean are called cyclones.

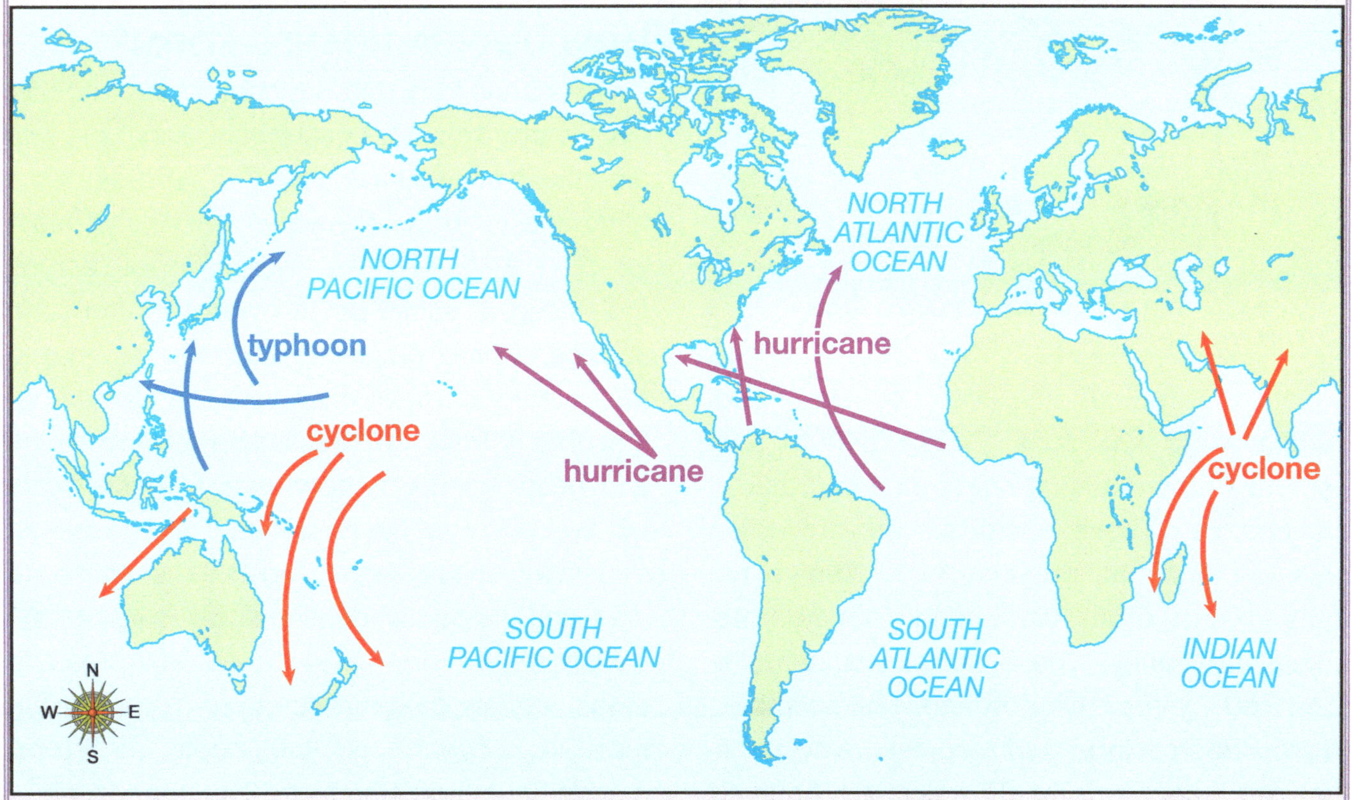

Fig. 7.1 Tropical cyclones have different names based on location.

Source: Mountain High Maps® Copyright © 1993-2020 Digital Wisdom®, Inc.

and typhoon is the location at which it forms. Hurricanes form in the Atlantic, Central North Pacific, and eastern North Pacific oceans. Tropical cyclones that form in the northwest Pacific are called typhoons. In the Indian and South Pacific Oceans, tropical cycles are called tropical cyclones.

Hurricane Harvey making landfall in Texas

Not all tropical cyclones develop into hurricanes or typhoons. Before a hurricane becomes a hurricane, it starts off as a tropical disturbance. These disturbances are low-pressure systems, but they have little to no rotation and often die before forming into actual hurricanes. These storm systems are classified by their wind speed. For example, tropical depressions are low-pressure systems that produce winds of 38 miles per hour or less. Stronger systems, known as tropical storms, are low-pressure areas that produce winds of 39–73 miles per hour. **Hurricanes** produce winds of 74 miles per hour or more. They are powerful, destructive storms. Hurricanes cause damage with ferociously strong winds, torrential rains, flooding, and even tornadoes. Hurricanes can destroy coastlines and entire cities, as well as create storm surges. A **storm surge** is a large wave created by a hurricane. This phenomenon occurs when water is pushed toward the shore by the force of the winds that are circling the storm.

Different areas of the world have different "hurricane seasons" in which the majority of hurricanes occur. For the United States and the rest of the Northern Hemisphere, the Atlantic hurricane season is from June 1 to November 30. According to the Tropical Cyclone Regional Specialized Meteorological Centre, over 130 tropical cyclones formed globally in 2019.

HOW DO HURRICANES FORM?

Hurricanes form when a cool air front hovers over warm, tropical ocean waters that are not directly at the equator (**Fig. 7.2**). In order for a hurricane to form, the water must be at least 26 °C (79 °F). As the storm moves across the ocean, it sucks up the warm air from the surface, which causes the colder air in the atmosphere to move underneath it. Remember that this is how low-pressure areas form. The sun heats up the surface water, which heats up the air. The air becomes less dense as molecules move farther apart. The warm air rises and the cooler, denser air takes its place.

Hurricanes form over these low-pressure areas. As low-pressure areas form due to the replacing of warm air with cooler air, storm clouds full of rain also form. The condensation of water into clouds releases heat. The heat warms the vapor as it condenses into colder air and begins to rise. The rising air is replaced by the warm surface air below. This process continues to occur and, in doing so, moves heat from the surface of the ocean to the atmosphere, creating a storm. The movement of heat and air also helps develop and organize the storm as the wind that circulates around the low-pressure point becomes similar to water flowing down a drain.

How are the strong winds that accompany hurricanes formed? As low-pressure areas form, winds converge at the ocean's surface and collide. This also pushes the warm, moist air up. This air joins with the other warm air that is rising from the surface of the ocean, reinforcing their strength and increasing wind speed. At the same time, the strong winds at the higher altitudes remove the hot air from the storm's low-pressure center. Since the hot air is being removed, it keeps the cycle going and the storm stays organized.

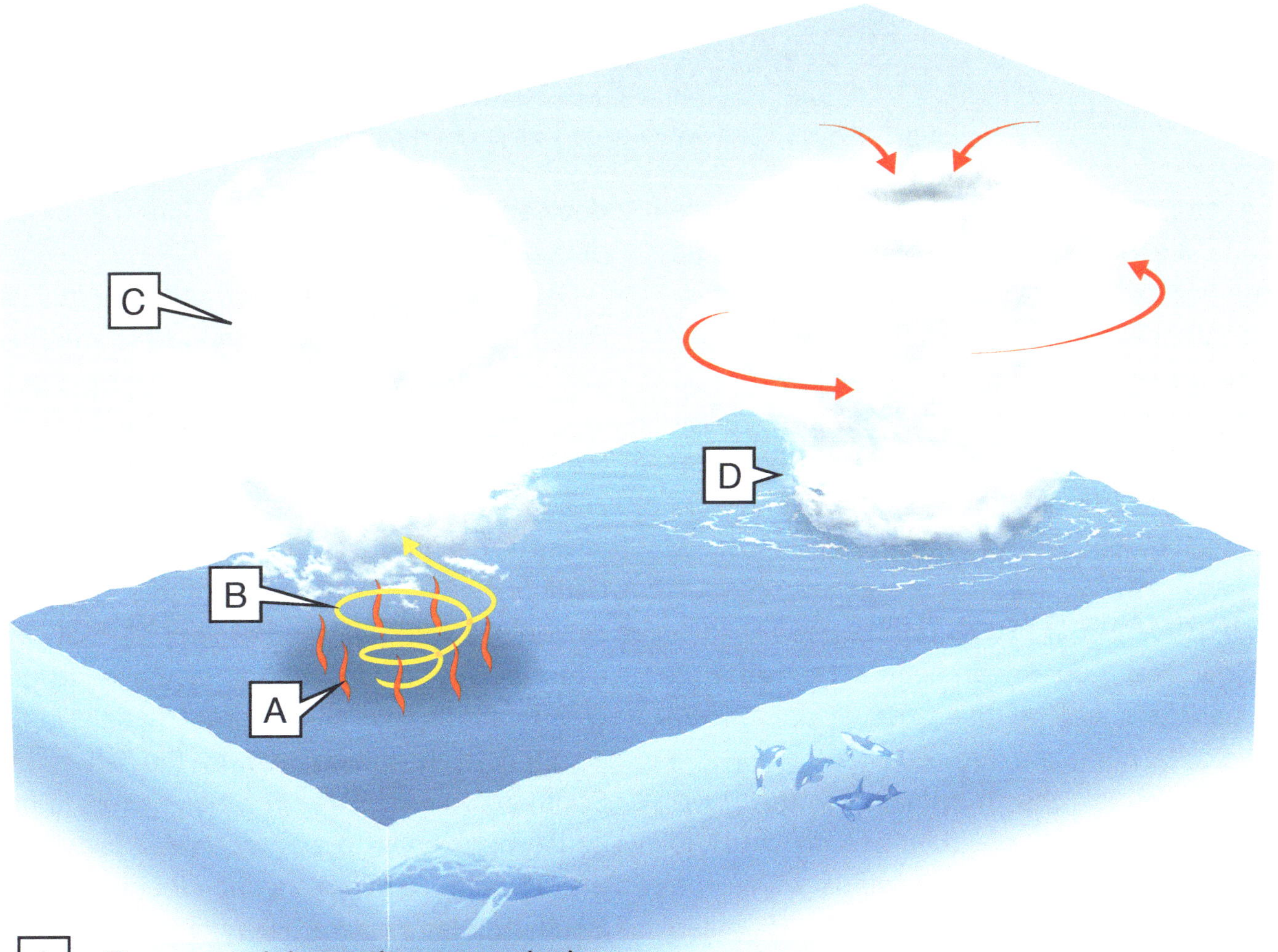

A > The warm air heats the ocean air above.

B > Rising warm air evaporates and starts to spin.

C > The air then cools and condenses to form a towering cumulonimbus cloud.

D > Intense low pressure sucks in air, causing very strong winds.

Fig. 7.2 Hurricane formation

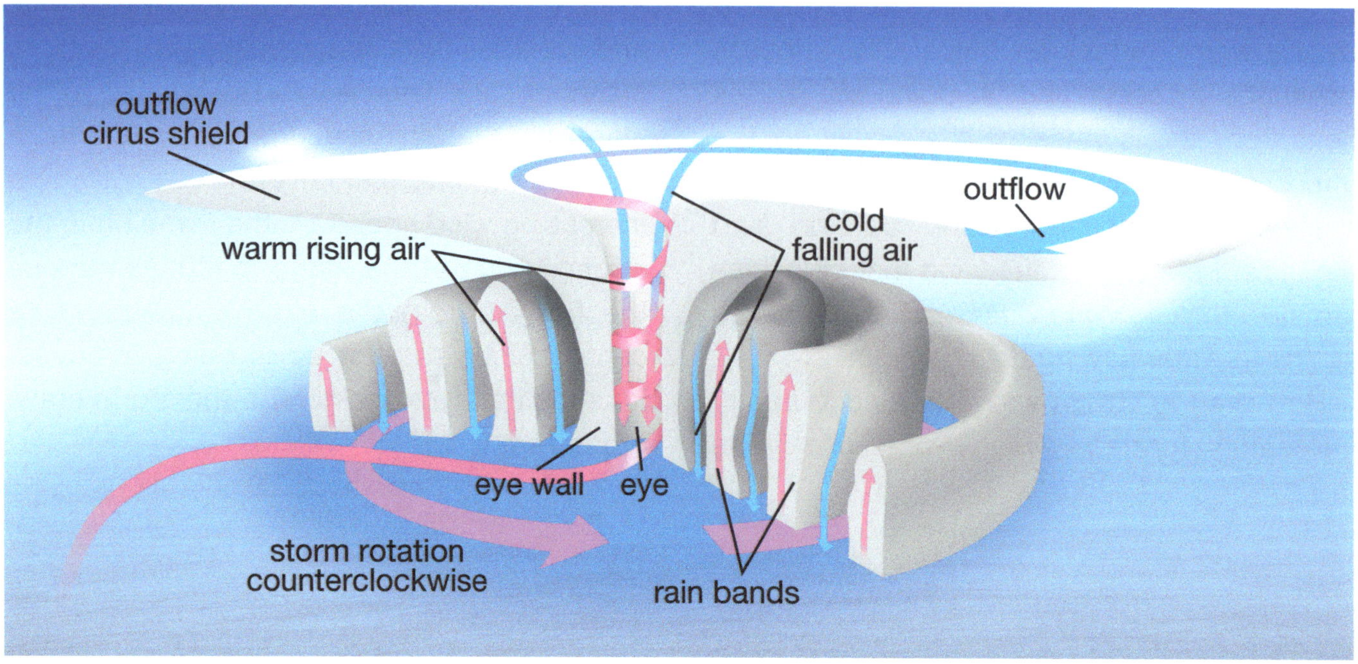

Fig. 7.3 Hurricane structure in the Northern Hemisphere

A storm can weaken if the high-altitude winds are not the same speed at the same time and place. However, the storm is further strengthened by even higher-altitude, high-pressure air, removing the warm air and then being sucked into the low-pressure center of the hurricane. In fact, hurricanes that form can release devastating damage with winds exceeding 185 miles per hour.

Hurricane Structure

Now that you know how a hurricane is formed let's look at the parts of an already developed hurricane (**Fig. 7.3**).

The eye of the hurricane is the low-pressure center of the storm and is relatively calm. In fact, the winds within the eye usually move no more than 15 miles per hour.

The eye wall of a hurricane is the area immediately surrounding the eye and is where the most violent wind occurs. This is because the eye wall is made up of a ring of thunderstorms. As the hurricane moves, changes to the eye wall can increase or decrease the winds, and therefore the impact of the storm.

Rain bands are also part of a hurricane. Rain bands are a string of thunderstorms that circulate outward from the eye of the storm. These thunderstorms feed and often intensify the hurricane through the evaporation and condensation cycle.

Once the parts come together to form a hurricane, the hurricane rotates because of the Coriolis effect. Remember that this effect is due to the rotation of the earth and causes winds to rotate toward the right in the Northern Hemisphere and to the left in the Southern Hemisphere. In other words, winds cause hurricanes to spin counterclockwise in the Northern Hemisphere and clockwise in the Southern Hemisphere. Hurricanes require the Coriolis effect in order to form. The path of a hurricane is also determined by this effect.

Hurricanes that form in the Atlantic Ocean often start close to Africa and move west using the trade winds. Once they meet the eastward winds coming from North America, they start to veer north. Hurricanes that form in the Pacific Ocean start in the warm waters of Central American and Mexican coastlines and move westward.

INTENSITY AND SCALE

The intensity of a hurricane is measured using the Saffir-Simpson Hurricane Wind Scale. This scale was developed in the 1970s to identify the potential levels of destruction a hurricane might cause. The Saffir-Simpson scale uses maximum sustained wind speed, pressure of the system, storm-surge height, and coastal destruction potential to assign a hurricane a level. The scale value is from Category 1, which has the lowest winds and least amount of damage potential, to Category 5 (**Table 7.1**).

Category 1 hurricanes will break branches off trees, damage power lines, and take the shingles off roofs. In 2012, when Hurricane Sandy hit the Jersey shore, it made landfall as a Category 1 hurricane. Category 2 hurricanes can cause siding damage, uprooted trees, and power loss for weeks. Category 3 hurricanes are the start of major hurricanes that will cause severe damage to homes, snap or uproot trees, block roadways, and cause power to be out for weeks.

In 2005, Hurricane Katrina reached Category 5 wind speeds but downgraded to a Category 4 by the time it hit land. Category 4 hurricanes have the second-worst storm value and will cause major structural damage to a home, uproot trees, block roadways, and cause power outages for months, leaving places uninhabitable for weeks or longer. In 2008, Hurricane Ike hit the Texas coast as a Category 4. Category 5 hurricanes are the most dangerous and are extremely rare, occurring about once every 100 years. These hurricanes will cause complete destruction by destroying homes, cause power outages for months, and leave areas uninhabitable for months.

Hurricane Andrew damage

TABLE 7.1 HURRICANE CATEGORY LABELS

Category	Wind Speed	Pressure	Storm Surge	Damage
1	74–95 mph	>980	4–5 feet	minimal
2	96–110 mph	979–965	6–8 feet	moderate
3	111–129 mph	964–945	9–12 feet	extensive
4	130–156 mph	944–921	13–17 feet	extreme
5	>157 mph	<920	>18 feet	catastrophic

Hurricane Andrew was a Category 5 hurricane that hit along the East Coast in 1992. It destroyed more than 63,000 homes, caused 20 billion dollars in damage, and killed 27 people. Hurricane Andrew reached winds of more than 160 mph and developed whirlwinds near the center of the storm that increased the destructive force of the storm.

NAMING HURRICANES

Hurricanes have been named since 1953 using a list of names created by the National Hurricane Center; however, they are now maintained by the World Meteorological Organization (WMO). It is easier to refer to hurricanes by names than using coordinates of longitude and latitude as they move across the ocean. Names also reduce confusion when there are multiple storms developing at the same time. There are six lists of names that are used in a rotation that cycles every six years. The only time the list is changed is if a deadly or costly storm occurs. In that case, the WMO meets to take the name off the list and replace it with another name. For example, the destruction of recent hurricanes such as Katrina, Sandy, and Ike have all warranted these names being retired.

REVIEW

Choose the best answers.

7.1) _____ The ___ is responsible for naming hurricanes.
- A. Richter scale
- B. World Meteorological Organization
- C. Saffir-Simpson scale
- D. Fujita scale

7.2) _____ The ___ is used to measure the damage of hurricanes from Category 1 to Category 5.
- A. Richter scale
- B. World Meteorological Organization
- C. Saffir-Simpson scale
- D. Fujita scale

7.3) _____ The calm center of a hurricane is called the ___.
- A. tropical storm
- B. tropical depression
- C. storm surge
- D. eye

7.4) _____ Low-pressure storm systems in the Atlantic Ocean with winds of 74 mph or more are called ___, or tropical cyclones.
- A. tropical storms
- B. tropical depressions
- C. storm surges
- D. hurricanes

7.5) _____ ___ have maximum winds of 38 mph.
- A. Tropical storms
- B. Tropical depressions
- C. Storm surges
- D. Hurricanes

7.6) _____ ___ have winds from 39–73 mph.

A. Tropical storms
B. Tropical depressions
C. Storm surges
D. Hurricanes

7.7) _____ A large wave created by a hurricane is a(n) ___.

A. tsunami
B. tropical depression
C. storm surge
D. eye

Order the steps of hurricane formation using the numbers 1–4.

7.8) _____ Rising warm air evaporates and starts to spin.

7.9) _____ Intense low pressure sucks in air, causing very strong winds.

7.10) _____ The warm air heats the ocean air above.

7.11) _____ The air then cools and condenses to form a towering cumulonimbus cloud.

Choose the best answers.

7.12) Choose TWO other names that are given to tropical cyclones based on location of formation.

☐ tsunamis
☐ monsoons
☐ typhoons
☐ tornadoes
☐ hurricanes

7.13) _____ What temperature does the water need to reach for a hurricane to form?

A. 100 °F
B. 0 °F
C. 95 °F
D. 79 °F
E. 32 °F

7.14) _____ What causes the storm surge that often follows a tropical cyclone making landfall?

A. global warming
B. low air pressure
C. winds pushing waves onshore
D. intense rainfall

Write T for True or F for False. If False, rewrite the statement to make it True.

7.15) _____ If you are planning a Caribbean vacation in August, you will need to watch for hurricanes.

____

____

7.16) _____ Hurricanes need cool water (at most 60 °F) to form.

____

____

7.17) _____ Winds circle counterclockwise north of the equator.

____

____

7.18) _____ Category 1 storms have the most damage potential.

____

____

Write the correct answer.

7.19) What role does the ocean play in the formation of hurricanes? _______________

____

____

CHECK	CORRECT	RECHECK
☐	☐	☐

8. CLIMATE

Climate, global warming, and climate change are important environmental issues. To understand how climate change and global warming work, we first need to understand how climate works. **Climate** is similar to weather, but it is also different. Climate describes the weather conditions in an area over a long period of time. It is the average weather over many years or decades. Different regions have different climates. For example, a tropical climate is warm and moist while a desert climate is hot and dry. While the weather for a day in the desert might include rain, the overall long-term weather patterns in the desert are hot and dry.

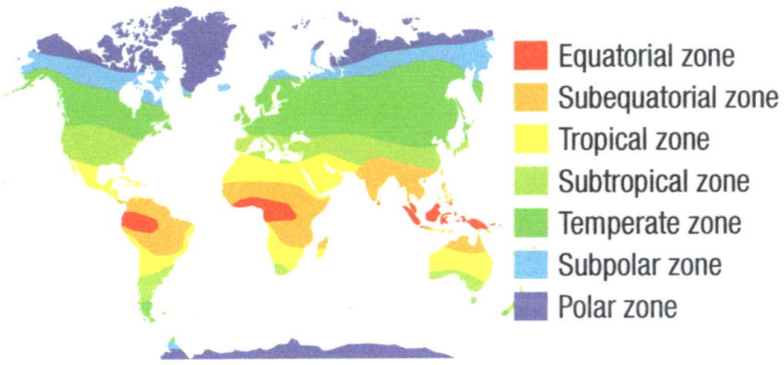

Climate distribution across the earth

Climate is affected by the same factors as weather (temperature, wind, rain, etc.). Let's discuss how some factors are climate specific.

- Sun – All weather starts with radiation from the sun entering the earth's atmosphere. Due to the earth's spherical shape and tilt, sunlight does not hit the surface evenly. This causes some places to receive more radiation and warmth than others. This is why it is warmer at the equator than at the poles.
The sun also varies slightly in the amount of radiation it gives off over long and short periods. Sunspots are dark spots on the sun's surface. The more sunspots present, the more radiation the sun gives off. There is typically an 11-year cycle during which the number of sunspots rises and falls. This is also referred to as the solar minimum and solar maximum.
- Ice – Ice reflects sunlight. This has an overall cooling effect on global climate.
- Greenhouse gases – Some gases called greenhouse gases in the atmosphere trap heat. The more greenhouse gases there are, the hotter the atmosphere gets.
- Topography – An area's topography includes land characteristics such as mountains, hills, lakes, and rivers. Altitude and mountain ranges are two topographic traits that affect climate. The higher the altitude, the cooler the temperature becomes. This is why snow can be found in the mountains of Hawaii and South Africa. Mountain ranges also influence wind patterns and precipitation.
- Volcanoes – Volcanoes influence climate because of the particles and gases they emit during an explosive eruption. These particles block out some of the sunlight that reaches the surface of the earth, having a cooling effect. Because of this, some researchers think volcanoes caused the Little Ice Age in the 1300s.
- Vegetation – When trees and other forms of vegetation cover a large region, they influence the climate. For example, the tropical rainforest in the Amazon covers an extensive area of land. All those trees

take in carbon dioxide (a greenhouse gas) and put water into the atmosphere. When forests are completely cleared out by deforestation, carbon is released back into the atmosphere. Regional temperatures rise because there is more carbon dioxide in the atmosphere and less water.

- Ocean – The ocean affects climate in many ways. One way is the water in the ocean is able to store heat, making a climate milder. Another way is the ocean moves heat around the globe through currents. If those currents were to change in a major way, an area's climate could become much cooler.

Many factors affect climate.

REVIEW

Choose the best answers.

8.1) _____ The weather conditions in an area over a long period of time is called ___.
- A. global warming
- B. climate

8.2) _____ What is weather?
- A. a change in air pressure, wind patterns, ocean currents, and ocean temperature
- B. the average atmospheric conditions over a short period of time
- C. the average atmospheric conditions over a long period of time
- D. a pattern of movement where less-salty water moves at the top of the ocean's surface and cold, salty water moves at the depths of the ocean

8.3) _____ How does the sun affect climate?
- A. All weather starts with radiation from the sun entering the earth's atmosphere.
- B. Volcanic particles can block out some of the sunlight that reaches the surface of the earth, having a cooling effect.
- C. Some gases in the atmosphere trap heat. The more greenhouse gases there are, the hotter the atmosphere gets.

8.4) _____ How do greenhouse gases affect climate?

A. Some gases in the atmosphere trap heat. The more greenhouse gases there are, the hotter the atmosphere gets.

B. All weather starts with radiation from the greenhouse gases entering the earth's atmosphere.

C. Volcanic particles can block out some of the sunlight that reaches the surface of the earth, having a cooling effect.

8.5) _____ How does topography affect climate?

A. Some gases in the atmosphere trap heat. The more greenhouse gases there are, the hotter the atmosphere gets.

B. Volcanic particles can block out some of the sunlight that reaches the surface of the earth, having a cooling effect.

C. The higher the altitude, the cooler the temperature becomes. Mountain ranges also influence wind patterns and precipitation.

8.6) _____ How do volcanoes affect climate?

A. Volcanic particles can block out some of the sunlight that reaches the surface of the earth, having a cooling effect.

B. The higher the altitude, the cooler the temperature becomes. Mountain ranges also influence wind patterns and precipitation.

C. Some gases in the atmosphere trap heat. The more greenhouse gases there are, the hotter the atmosphere gets.

8.7) _____ How does vegetation affect climate?

A. Trees and vegetation keep temperatures lower by taking in carbon dioxide and transpiring water into the atmosphere.

B. Some gases in the atmosphere trap heat. The more greenhouse gases there are, the hotter the atmosphere gets.

C. The higher the altitude, the cooler the temperature becomes. Mountain ranges also influence wind patterns and precipitation.

8.8) _____ How does the ocean affect climate?

A. The ocean stores heat, moves heat through currents, and stores carbon dioxide, keeping things from getting too warm.

B. Trees and vegetation keep temperature lower by taking in carbon dioxide and transpiring water into the atmosphere.

C. Some gases in the atmosphere trap heat. The more greenhouse gases there are, the hotter the atmosphere gets.

Write T for True or F for False. If False, rewrite the statement to make it True.

8.9) _____ Climate and weather are the same.

____

____

8.10) _____ Studying the climate is simple.

____

____

Write whether each effect is C for Cooling or W for Warming the climate.

8.11) _____ ice reflecting sunlight

8.12) _____ erupting volcanoes

8.13) _____ the location of the equator

8.14) _____ high altitude of a mountain

8.15) _____ removal of a tropical rainforest

Climate Change

Some of the "hottest" environmental topics today are global warming and climate change. Global warming and climate change are two terms that are sometimes used interchangeably, but they are actually different. **Global warming** is the rising of the overall average global temperature. **Climate change** is a long-term change in a region's weather conditions such as precipitation and temperature. Global warming may cause climate change. A region's climate change might not end up being warmer but colder. This is due to the fact that climate and weather are intertwined in a complicated system. A warming in one spot may result in cooling in another area.

Evidence of Climate Change

How do scientists know that global warming is happening? Many people rely on anecdotal evidence (observations of what they are experiencing) to form their opinions on global warming and climate change. Scientists look at long-term data trends as evidence of climate change.

Much of the theory of global warming revolves around the greenhouse effect. The **greenhouse effect** is the trapping of heat that reradiates from the surface of the earth. The greenhouse effect is good. Without it, the earth would be too cold for survival. However, as with so many other things in life, too much of a good thing can be bad. First, let's see how the greenhouse effect works (**Fig. 8.1**):

1 – Sunlight radiates through the atmosphere and hits the earth's surface.

2 – The earth's surface reradiates the heat outward. This is also called infrared energy.

3 – Greenhouse gases in the atmosphere trap some of the reradiated heat and keep the heat closer to the surface of the earth. This is much like the inside of a car on a sunny day. The sunlight radiates through the glass. The light energy hits the seats and dashboard

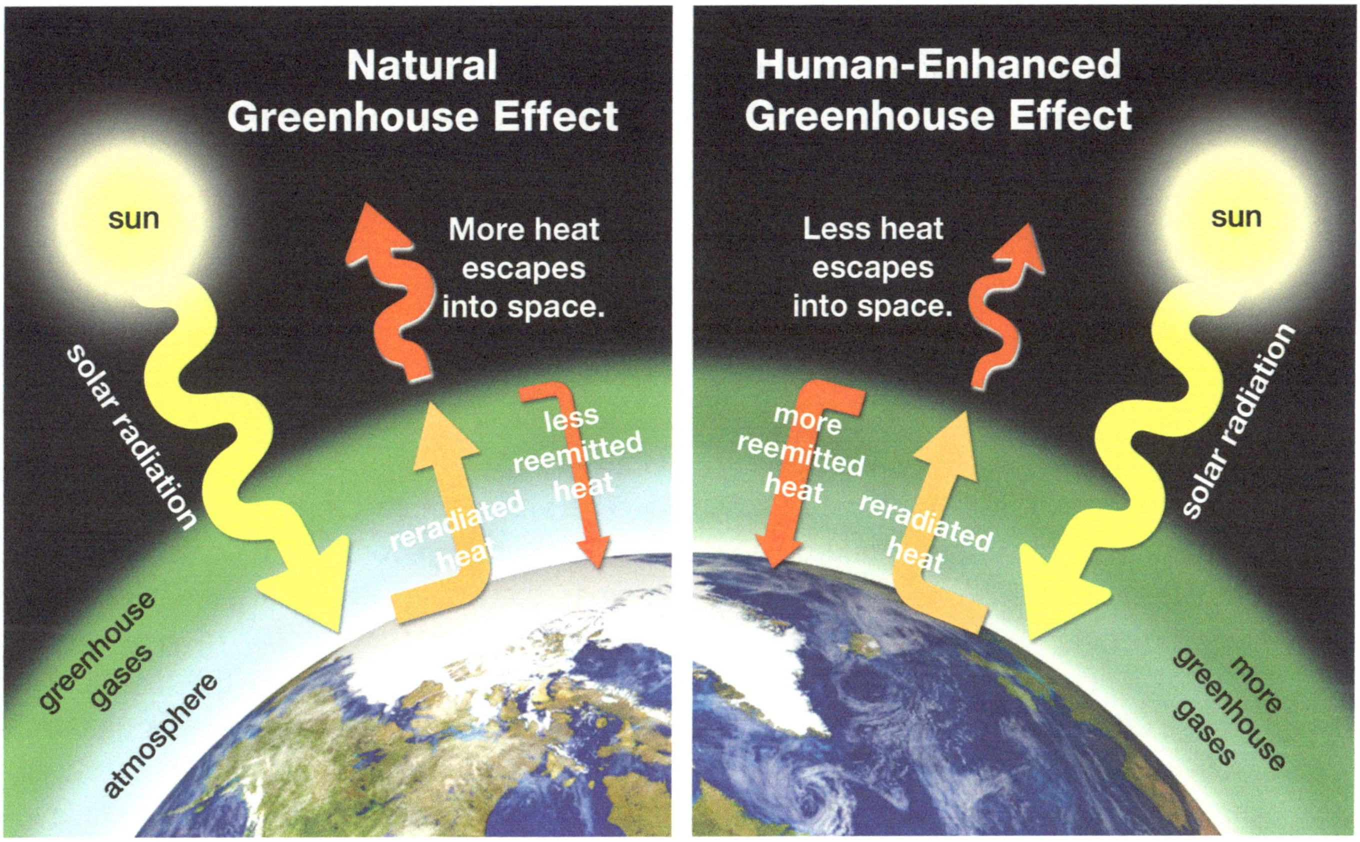

Fig. 8.1 The greenhouse effect

Fig. 8.2 Sources of greenhouse gases

and converts to heat energy through conduction. However, much of the heat energy cannot escape through the glass. The inside of the car then becomes much warmer. The car's glass is like greenhouse gases in the atmosphere.

The most common greenhouse gases are carbon dioxide, methane, nitrous oxide, and water vapor. Sources of greenhouse gases, both natural and human enhanced, are shown in **Fig. 8.2.** Many greenhouse gases remain in the atmosphere for tens to hundreds of years.

Their warming effects on the climate can last for very long periods of time and impact current and future generations.

If more greenhouse gases are added to the atmosphere, they can trap more reradiated heat. Trapping more heat leads to warmer surface temperatures and an increase in the carbon dioxide levels in our atmosphere. This rise in surface temperatures and carbon dioxide levels has been linked to the following climate phenomena.

Rising Ocean Levels

The rate of global sea level rise is increasing. Global average sea levels rose about 21 centimeters, or 8 inches, between 1880 and 2020. This rate is twice that of the century before. As sea levels rise, they cause an increase in the intensity of storm surges, flooding, coastal erosion, and the loss of habitat for fish, birds, and plant life.

Warmer Oceans

Besides surface temperatures increasing, ocean temperatures have risen as well. Rising ocean temperatures threaten natural ecosystems and cause changes in ocean health and biochemistry.

Melting Ice

Due to an increase in global temperatures, sea level rise is accelerating rapidly. Ice sheets, icecaps, glaciers, and sea ice are melting at historically high rates, releasing water into the ocean. This influx of water causes catastrophic floods, landslides, habitat loss, and other environmental issues. Many animal species, such as polar bears, seals, foxes, wolves, and penguins rely on icy habitats to survive and are threatened by global melting.

Extreme Weather

There have been a number of record-high temperatures recorded in different cities across the United States in recent years. At the same time, the number of record lows has been decreasing since the 1950s. The number of intense rainfall events has also increased. Although the number of severe storms such as hurricanes has not increased, the number of extremely intense occurrences of these storms has increased. For example, Typhoon Haiyan was one of the strongest and deadliest tropical storms to make landfall in the Philippines in November of 2013. With a wind speed of 195 mph and rainfall of 28.19 cm, this typhoon claimed more than 6,300 lives and caused billions of dollars in damage.

Ocean Acidification

Since the Industrial Revolution, the surface of the ocean has increased its acidity by 30%. The acidity increases as the oceans take in more atmospheric carbon dioxide. Carbon dioxide enters surface ocean water at a rate of about 2 billion tons per year.

Choose the best answers.

8.16) _____ The gradual increase in the earth's temperature is called ___.

A. global warming

B. climate

8.17) Choose the FOUR most common greenhouse gases.

☐ carbon dioxide
☐ methane
☐ water vapor
☐ nitrous oxide
☐ helium
☐ oxygen

8.18) _____ Why is the greenhouse effect good?

A. Without the greenhouse effect, the earth would be too cold for survival.
B. Without the greenhouse effect, the earth would be too oxygen deprived for survival.
C. Without the greenhouse effect, the earth would be too barren for survival.
D. Without the greenhouse effect, the earth would be too hot for survival.

Write T for True or F for False. If False, rewrite the statement to make it True.

8.19) _____ Global warming is only caused by people.

____

____

CHECK ☐ CORRECT ☐ RECHECK ☐

9. CLIMATE CHANGE: NATURAL VS. HUMAN IMPACT

As you learned in Lesson 8, there is clear evidence that global temperatures are increasing and that the cause is a result of both natural and human-enhanced events. Throughout its existence, the earth has gone through several warming and cooling periods. These periods happened long before humans made an impact on the environment. There are many environmental factors that have triggered these warming and cooling periods, influencing climate. These factors include changes in the amount of greenhouse gases in the air, changes in solar output, and changes in ice covering the globe. We will discuss other factors affecting climate change throughout this Lesson.

IS IT CAUSED BY NATURE?

The warming we see today may be a result of natural events such as changes in the sun's intensity and even slight changes in the earth's orbit. Other natural events that can trigger climate change are volcanism, meteor impacts, and abrupt changes in ocean currents.

Volcanoes

Ash carried into the atmosphere can spread and block the sun's rays, resulting in a cooler earth.

When volcanoes erupt, the soot, ash, dust, and aerosols they emit get added to the atmosphere. Large eruptions of these airborne particles have the potential to travel hundreds to thousands of kilometers away from the source and affect climates in places on the other side of the globe. The sun's rays can be blocked by the ash thrown into the sky during an eruption, creating a cooling effect.

Meteors

When a meteor impacts the earth's atmosphere from space, the impact it makes when it hits the surface can cause global changes. The friction created by the ash, dust, and greenhouse gases ejected from this massive space rock can cause temporary heating of the earth's surface. This rise in temperature can be several hundred degrees Celsius. This intense heating is usually followed by a significant drop in temperatures for months or even years at a time. The atmospheric particles emitted cause a partial blockage of sunlight. Many believe that this type of astronomical event triggered the start of the last ice age on the earth, leading to the extinction of the dinosaurs that once roamed our planet.

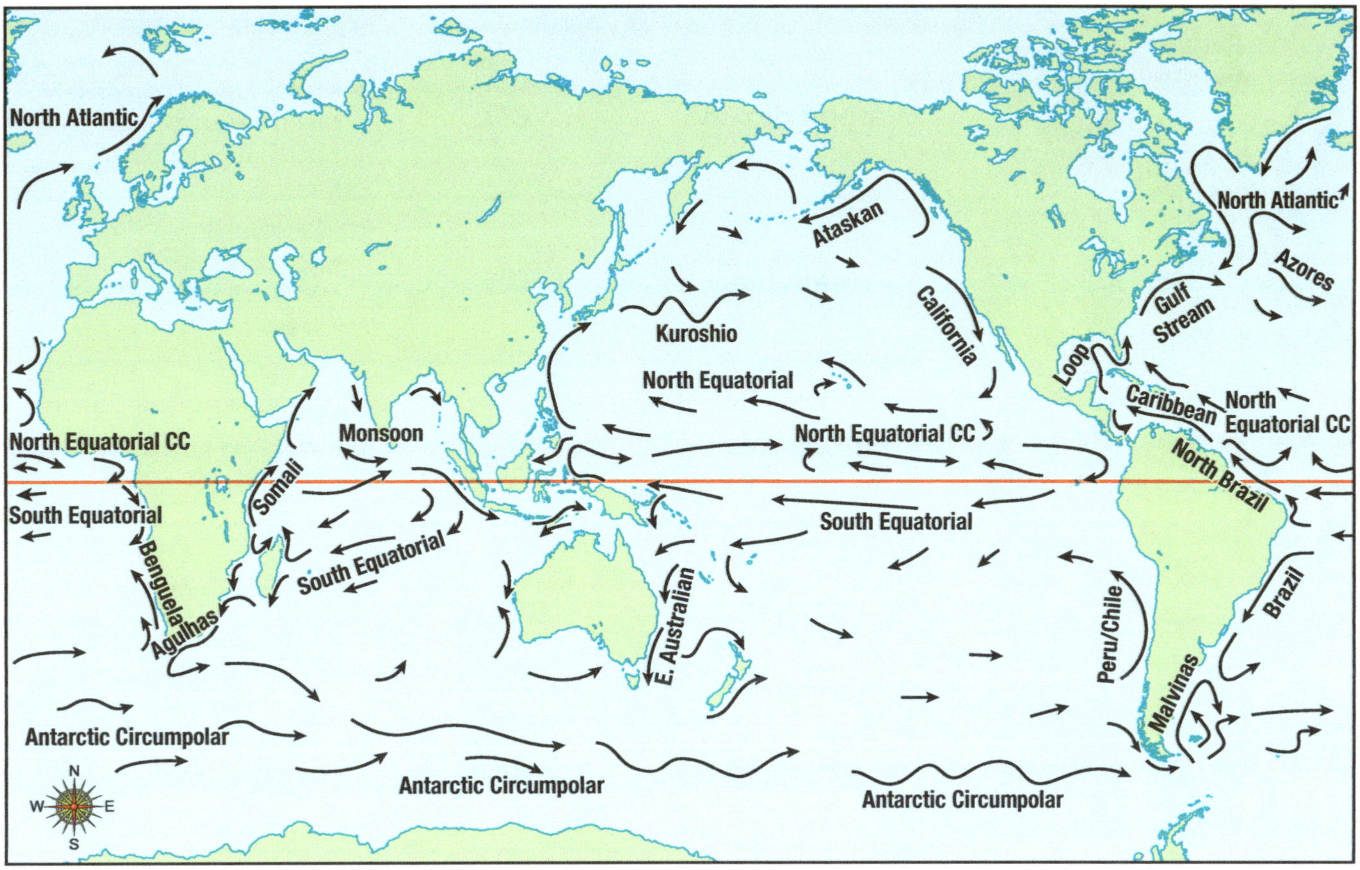

Fig. 9.1 Ocean currents influence climate and weather patterns around the world.

Source: Mountain High Maps® Copyright © 1993-2020 Digital Wisdom®, Inc.

Ocean Currents

Shifts in ocean currents can also have a significant impact on the global climate (**Fig. 9.1**). Driven by fast winds, ocean currents carry heat around the earth. As the seas take in more heat from the atmosphere, sea surface temperatures rise. This rise in temperature causes the ocean circulation patterns that transport warm and cold water around the globe to change. Different regions may become warmer or colder depending on how these currents change their course. Increases in sea surface temperature can increase the amount of water vapor that forms over the seas. This increases the amount of greenhouse gases. Warmer waters mean that oceans cannot take in as much carbon dioxide from the atmosphere, leading to changes in local weather patterns.

Because ocean currents are responsible for moving heat away from the equator and towards the poles, they help maintain the natural order and balance of the climate. Global weather patterns would be disrupted if ocean currents abruptly changed. For instance, temperatures would cool if ocean currents slowed down. Cold ocean currents contribute to the formation of coastal deserts. Although the winds that blow over these deserts are from across the ocean, they do not bring rain. A lack of rainfall in coastal areas can lead to habitat destruction, biodiversity loss, soil erosion, and a higher risk of wildfires.

Warmer ocean currents speed up the rate of evaporation, producing more rainfall in coastal regions. Increases in precipitation could increase the intensity of storm surges and flooding in surrounding areas.

MAIN GREENHOUSE GASES

	Greenhouse Gas	Chemical Formula	Anthropogenic Source
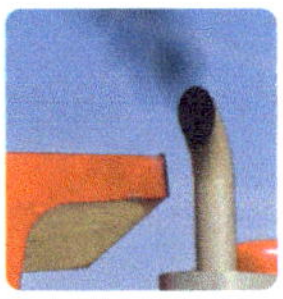	carbon dioxide	CO_2	• fossil fuel combustion • land use conversion • cement production
	methane	CH_4	• fossil fuels • rice paddies • waste dumps • livestock
	nitrous oxide	N_2O	• fertilizer • industrial processes • combustion
	CFC-12	CCl_2F_2	• liquid coolants • foams
	HCFC-22	$CHClF_2$	• liquid coolants
	perfluoromethane	CF_4	• aluminum production
	sulphur hexafluoride	SF_6	• electric fluid

Fig. 9.2 Anthropogenic greenhouse gases

IS IT CAUSED BY HUMANS?

Natural changes alone do not account for current temperature increases. The majority of scientists believe that the current global warming trend is caused by humans. This belief is known as **anthropogenic** global warming. Scientific data shows that the increase in atmospheric carbon dioxide comes mostly from humans. Since the Industrial Revolution began in 1750, human activities have contributed significantly to the emission of greenhouse gases. The burning of fossil fuels, such as coal, oil, and gas, has been the largest contributor to global climate change to date.

Fig. 9.2 lists more information about human-enhanced greenhouse gases.

Deforestation

Deforestation and forest degradation caused by the temporary or permanent removal of forests have also increased the amount of carbon dioxide in the atmosphere. Having fewer trees to remove carbon dioxide from the air contributes to climate change. Also, deforestation can lead to an increase in temperature, as more carbon dioxide is left in the atmosphere and less water vapor is transpired. Transpiration, or the releasing of water into the atmosphere by plants, has a cooling effect. This is one reason why it feels cooler to sit under the shade of a tree than under the shade of a vinyl canopy. Without forests, a region's climate can become warmer and drier.

The deforestation and overuse of tree products is a modern issue with long-term implications on the environment.

Urbanization

As more people move from rural areas to cities seeking economic opportunity, urbanization (the population shift from rural to urban areas) is becoming another main cause of climate change. Over half of the world's population lives in cities, and the numbers are steadily climbing. Cities use a large amount of the world's energy supply and produce nearly three-quarters of the greenhouse gas emissions. These gases trap heat and result in the warming of the earth. The burning of fossil fuels and industrial waste are just two examples of human activities in cities that contribute to the environment's pollution problem and threaten both human and animal health. The growth of cities can impact the earth's system in a number of ways. For example, in urbanized areas, the heat from the sun is absorbed and reemitted by buildings, industrial plants, vehicles, and air conditioners. This human-generated waste heat can contribute to the heat island effect. Heat islands increase air pollution levels, reduce nighttime cooling, and cause higher daytime temperatures.

An increase in global surface temperatures increases the possibility of more intense storms, resulting in coastal flooding and habitat destruction. As certain regions warm, some tropical insects will also migrate with the warm weather and spread disease. Some areas will see an increase in food production while others will see a decrease due to increased drought. As cities expand, they become a major driver of environmental change. Landscapes are transformed, affecting ecosystems and biodiversity in the surrounding areas.

Tipping Point

It is hard to imagine the future climate of the earth. Making predictions about a week of weather can often prove difficult, let alone trying to forecast 25, 50, and 100 years into the future. However, scientists are trying to better understand some of the changes that may occur so that we can prepare for any changes that may be coming.

Many scientists believe that we could reach a tipping point. A **tipping point** is a point at which a series of small changes triggers a larger, more important change. A climate tipping point is the point at which severe climate disruption cannot be reversed. Scientists have no way of knowing what the tipping point is, if one exists at all, until we pass it. If a tipping point is reached, the earth will experience drastic, rapid changes that will be irreversible for an indefinite period of time.

Even without a tipping point, many greenhouse gases remain in the atmosphere for long periods of time. For example, carbon dioxide (CO_2) remains in the atmosphere for up to 200 years or more. Even if people drastically reduce their greenhouse gas emissions, the change will be ineffective for hundreds of years. Because the ocean takes decades to respond to warming, temperatures may continue to rise even after greenhouse gas levels stabilize.

If this current warming trend is caused by humans, as the majority of scientists believe, then the good news is that humans can possibly slow it down through prevention and cleanup strategies. Many prevention strategies focus on shifting away from fossil fuels, especially coal, to "cleaner" and more renewable forms of energy like wind and solar power and reducing deforestation. A few common cleanup strategies include removing carbon from the atmosphere by planting more trees or capturing and storing it in the soil or oceans.

Making predictions about the future climate is not an exact science, but it is still important. As more data becomes available and climate models become more accurate, climate predictions, just like weather forecasts, will continue to change. The more we know, the sooner we can take action and make preparations.

REVIEW

Choose the best answers.

9.1) _____ Define the term anthropogenic.

A. the study of climate through all of the earth's history
B. a long-term change in a region's precipitation and temperature
C. the rising of the overall global temperature
D. caused by humans

9.2) _____ What is anthropogenic global warming?

A. the trapping of heat that reradiates from the surface of the earth
B. the belief that humans are causing global warming
C. the rising of the overall global temperature
D. a long-term change in a region's precipitation and temperature

9.3) _____ How could added greenhouse gases from the burning of fossil fuels affect global temperatures?

A. More gases prevent reradiated heat. The trapping of no heat leads to a dramatic decrease in global temperatures.
B. More gases trap extreme amounts of reradiated heat. The trapping of more heat may lead to the earth becoming too hot for humans to live.
C. More gases trap less reradiated heat. The trapping of less heat may lead to a decrease in global temperatures.
D. More gases trap more reradiated heat. The trapping of more heat may lead to an increase in global temperatures.

9.4) _____ Define the term tipping point.

A. the temperature at which a liquid boils and turns to vapor
B. the temperature at which a liquid turns into a solid when cooled
C. the point at which a series of small changes becomes significant enough to cause a larger, more important change
D. the time at which a volcano is predicted to erupt

9.5) Choose THREE natural causes for climate change.

☐ industry
☐ slight changes in the earth's orbit
☐ changes in solar output
☐ volcanic activity
☐ burning fossil fuels

9.6) Choose THREE anthropogenic causes for climate change.

☐ industry
☐ slight changes in the earth's orbit
☐ deforestation
☐ volcanic activity
☐ urbanization

Write the correct answers.

9.7) Mark each cause as Natural or Anthropogenic

Cause	Natural	Anthropogenic
Industry		
Slight changes in the earth's orbit		
Changes in solar output		
Burning fossil fuels		
Deforestation		
Volcanic activity		
Urbanization		
Ocean currents		

Choose the best answers.

9.8) _____ Which activity is the largest contributor of greenhouse gases?

A. deforestation
B. burning of fossil fuels
C. volcanoes
D. agriculture

9.9) _____ How long does carbon dioxide remain in the atmosphere?

A. 5–7 days
B. 12 months
C. 5–10 years
D. 200 years or more

9.10) _____ Deforestation contributes to greenhouse effect by ___.

A. increasing carbon dioxide concentrations in the atmosphere, which increases global temperatures
B. increasing carbon dioxide concentrations in the atmosphere, which decreases global temperatures
C. decreasing carbon dioxide concentrations in the atmosphere, which increases global temperatures
D. decreasing carbon dioxide concentration in the atmosphere, which decreases global temperatures

9.11) _____ What effect do abrupt changes in ocean currents have on climate?

A. Ocean currents can carry warm water, which can raise temperatures and cause more precipitation.
B. Ocean currents can carry warm water, which can decrease temperatures and cause less precipitation.
C. Ocean currents can carry cold water, which can raise temperatures and cause less precipitation.
D. Ocean currents can carry cold water, which can decrease temperatures and cause more precipitation.

9.12) _____ The majority of scientists agree that climate change is real and is mainly caused by nature.

A. True
B. False

9.13) _____ The biggest anthropogenic factor influencing the climate is the emission of greenhouse gases caused by urbanization.

A. True
B. False

9.14) _____ Carbon dioxide released into the atmosphere by meteor impacts immediately cools and then heats the earth's surface by hundreds of degrees Celsius.

A. True
B. False

CHECK ☐ CORRECT ☐ RECHECK ☐

ACTIVITY: Climate Change

Write a scientific argument that uses evidence and reasoning to support your claims about the potential causes and consequences of global warming and climate change.

PART 1: CLAIM, EVIDENCE, AND REASONING GRAPHIC ORGANIZER

- Create a claim, evidence, and reasoning statement about climate change and fill out the graphic organizer.
- Research and explore resources such as books, articles, websites, and personal observations to gather evidence.
- Cite at least three sources of scientific studies with details about who conducted the study, who funded the research, and where the information was published or presented.

PART 2: COUNTERARGUMENT

- Rebuttal – State a counterargument to your claim. How could the claim be argued against? Are there other opinions and evidence different from the claim?
- Discussion – Share your argument and practice your rebuttal and counterargument with a partner.
- Reflection – Share key points from your discussion with the class.

RUBRIC

Criteria	Exemplary 4	Accomplished 3	Developing 2	Beginning 1	Score
Claim: An inference or a statement that answers the question.	The student makes an accurate and complete claim.	The student makes an accurate but incomplete claim.	The student makes an inaccurate claim.	The student does not make a claim.	____
Evidence: Scientific data that supports the claim. Cites at least 3 sources of scientific studies.	The student uses scientific data that supports the claim and cites at least 3 sources of scientific studies.	The student uses scientific data that supports the claim and cites at least 2 sources of scientific studies.	The student uses scientific data that supports the claim and cites at least 1 source of scientific studies.	The student does not use scientific data that supports the claim.	____
Reasoning: A justification that connects the evidence to the claim.	The student provides reasoning that appropriately connects each piece of evidence to the claim and uses scientific principles to explain why the evidence supports the claim.	The student provides appropriate but incomplete reasoning. Each piece of evidence is not supported by a line of reasoning.	The student provides inappropriate reasoning.	The student does not provide reasoning.	____
Rebuttal: Provides an alternative explanation, counter-evidence, and reasoning for why the original argument is not appropriate.	The student gives an alternative explanation and provides appropriate and sufficient counter-evidence and reasoning when making a rebuttal.	The student gives an alternative explanation and provides appropriate counterevidence but insufficient reasoning when making a rebuttal.	The student gives an alternative explanation but does not provide appropriate or sufficient counter-evidence or reasoning when making a rebuttal.	The student does not provide a rebuttal or makes an inaccurate rebuttal.	____

If not attempted, score = 0.   Total _____/16

TEACHER CHECK ☐

MY SCORE

QUIZ 3

TEACHER INITIAL

(Each answer, 4 points)

Match the hurricane-related words with the descriptions.

3.01) _____ storm surge

3.02) _____ eye

3.03) _____ hurricane

3.04) _____ Saffir-Simpson

3.05) _____ tropical depression

3.06) _____ tropical storm

A. the scale used to measure hurricane damage

B. a storm with maximum winds of 39–73 mph

C. a storm with maximum winds of 38 mph

D. the calm center of a hurricane

E. a large wave created by a hurricane

F. a low-pressure storm system that produces winds of 74 mph or more

Match the climate-related words with the descriptions.

3.07) _____ global warming

3.08) _____ weather

3.09) _____ greenhouse effect

3.010) _____ climate

3.011) _____ climate change

3.012) _____ anthropogenic

A. a gradual increase in the earth's temperature

B. the weather conditions in an area over a long period of time

C. a change caused by humans

D. the state of the atmosphere at a given time and place

E. the trapping of heat that reradiates from the surface of the earth

F. a long-term change in a region's precipitation and temperature

Write N for Natural or A for Anthropogenic.

3.013) _____ burning of fossil fuels

3.014) _____ volcanic activity

3.015) _____ slight changes in the earth's orbit

3.016) _____ changes in solar output

3.017) _____ abrupt changes in ocean currents

3.018) _____ deforestation

3.019) _____ urbanization

Choose the best answers.

3.020) _____ Which of the following is the first step in hurricane formation?

A. Rising warm air evaporates and starts to spin.

B. Intense low pressure sucks in air, causing very strong winds.

C. The warm air heats the ocean air above.

D. The air then cools and condenses to form a towering cumulonimbus cloud.

3.021) _____ An erupting volcano has a ___ effect on climate.

A. warming

B. cooling

3.022) _____ Removing tropical rainforests has a ___ effect on climate.

A. warming

B. cooling

3.023) _____ Ice reflecting sunlight has a ___ effect on climate.

A. warming

B. cooling

3.024) _____ Which temperature does the ocean water need to be for hurricanes to form?

A. 79 °C (26 °F)

B. 10 °C (50 °F)

C. 100 °C (212 °F)

D. 26 °C (79 °F)

3.025) Which TWO names are given to tropical cyclones based on the location of their formation?

☐ tsunamis

☐ typhoons

☐ tornadoes

☐ hurricanes

☐ monsoons

CHECK ____ CORRECT ____ RECHECK ____

STOP and prepare for the Unit Practice Test.

- Review the Objectives and Vocabulary for each Section.
- Reread each Lesson and its corresponding questions.
- Relearn each Lesson that you still do not understand.
- Review the Quizzes.

MY SCORE

PRACTICE TEST

TEACHER INITIAL

(Each answer, 3 points)

Fill in the blanks using words from the boxes below.

moon	Coriolis effect	polar air mass	jet streams
convection current	upwelling	ocean currents	deep current
sun			

1) A(n) ______________________________ is not as cold as arctic air and begins over high latitudes over land or sea.

2) The ______________________________ are high speed winds that blow from west to east.

3) The curving path an object takes due to the earth's rotation is called the ______________________________.

4) The ____________________ drives convection within the earth's atmosphere and oceans.

5) Water that moves due to the difference in density is the ______________________________.

6) When observing the ocean, the movements are the ____________________________.

7) Nutrients are brought to the surface by a(n) ____________________.

global warming	surface current	greenhouse effect	cold front
pressure belt	meteorologist	warm front	

8) A ______________________________ is a current that only goes about 100 meters deep into the ocean.

9) A ______________________________ causes temperatures to fall and thunderstorms to develop.

10) A ____________________ studies weather and the earth's atmosphere to predict weather forecasts.

11) A ______________________________ is a boundary that forms between latitudes.

12) ______________________________ is the rising of the overall global temperature.

13) The ______________________________ occurs when gases in the earth's atmosphere trap the sun's heat.

continental air mass	California Current	El Niño
tropical depression	tropical air mass	Gulf Stream
high-pressure system	low-pressure system	La Niña
maritime air mass	arctic air mass	hurricanes
updraft	storm surge	

14) Florida has mild winter temperatures due to the ______________________.

15) Temperatures in the Pacific Ocean are below average because of ______________________.

16) Temperatures in the Pacific Ocean are higher than average because of ______________________.

17) ______________________ need ocean water to reach 26 °C (79 °F) in order to form.

18) A(n) ______________________ is a large wave created by a hurricane.

19) As warm air rises, the ______________________ allows the winds in the storm to spin.

20) A(n) ______________________ is cool, dry air that allows for sunny days.

21) A(n) ______________________ brings warm air up from low latitudes over land or sea.

22) A(n) ______________________ starts in the ocean and has a high moisture count.

23) A(n) ______________________ brings cold air from the arctic region.

24) A(n) ______________________ is warm, moist air that allows for storms.

25) A(n) ______________________ starts over the land and contains cold, dry air.

26) A(n) ______________________ is a storm with maximum winds of 38 mph.

Match the words with the descriptions.

27) _____ air pressure

28) _____ atmospheres

29) _____ humidity

30) _____ precipitation

31) _____ temperature

32) _____ wind

A. rain, sleet, hail, or snow

B. the air movement in the atmosphere

C. a unit of measurement for air pressure

D. the measure of how hot or cold the atmosphere is

E. the amount of water vapor in the air

F. the weight of the air on the surface of the earth

Choose the best answer using the map.

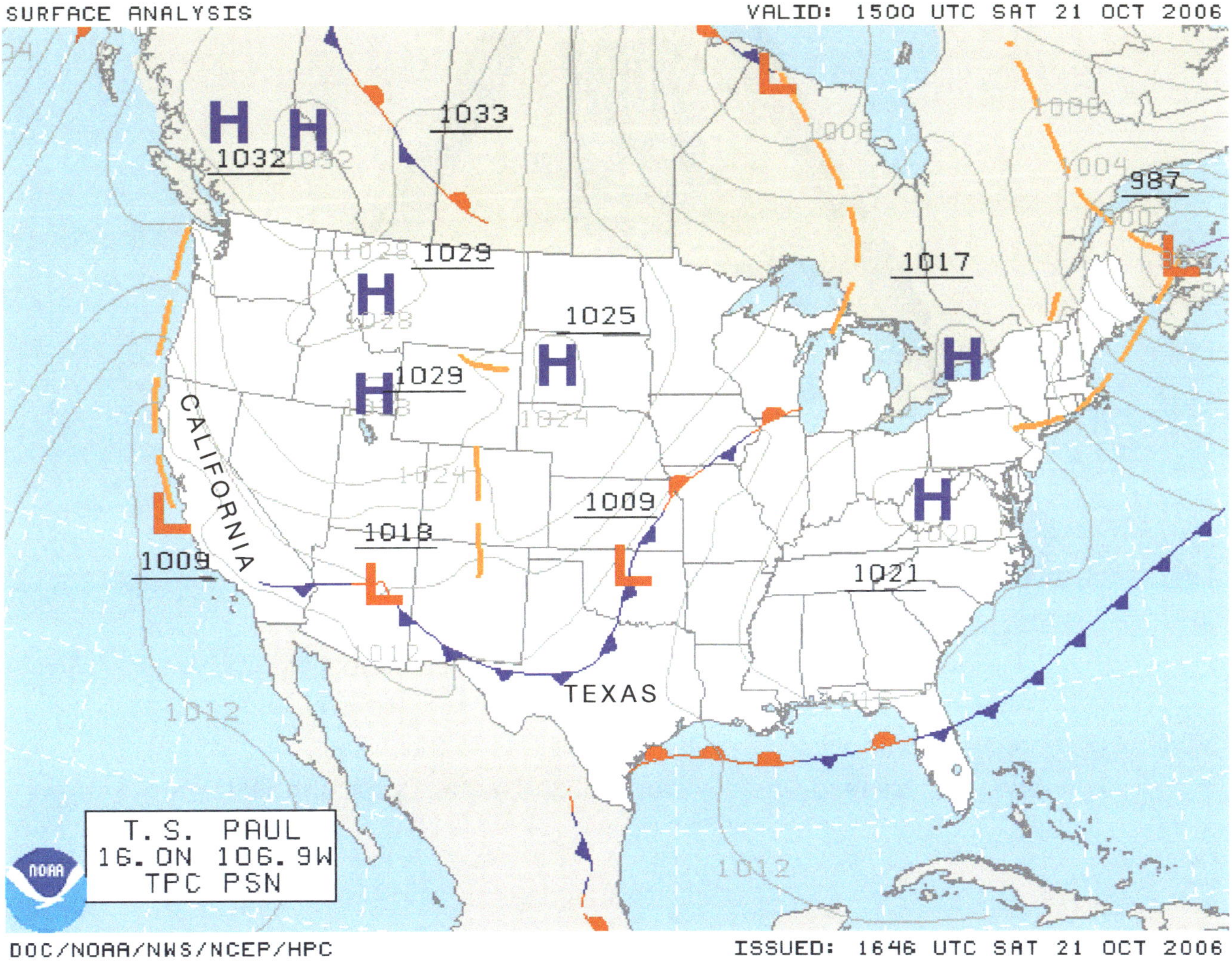

33) _____ Which type of front can be found in Texas?

A. high front

B. occluded front

C. warm front

D. cold front

CHECK ☐ CORRECT ☐ RECHECK ☐

You must now prepare for the Unit Test.

- Review the Objectives and Vocabulary for each Section.
- Reread each Lesson and its corresponding questions.
- Review and study the Quizzes and Unit Practice Test.

When you are ready, turn in your Unit and request your Unit Test.